The ASQ Certified Supplier Quality Professional Study Guide

The ASQ Certified Supplier Quality Professional Study Guide

Second Edition

Stephanie Parker, Editor

ASQExcellence
Milwaukee, Wisconsin

The ASQ Certified Supplier Quality Professional Study Guide, Second Edition
Produced and distributed by Quality Press, ASQ, Milwaukee, WI

Publisher's Cataloging-in-Publication Data

Names: Parker, Stephanie L., 1973-, editor.
Title: ASQ certified supplier quality professional study guide , second edition /
Stephanie Parker, editor.
Description: Milwaukee, WI: ASQExcellence, 2025.
Identifiers: LCCN: 2025934147 | ISBN: 9781636941585 (paperback) | 9781636941592 (PDF) |
9781636941608 (epub)
Subjects: LCSH Industrial procurement—Quality control—Study guides. |
Materials management—Quality control—Study guides. | Quality control—
Standards—Study guides. | BISAC BUSINESS & ECONOMICS / Logistics &
Supply Chain | BUSINESS & ECONOMICS / Quality Control
Classification: LCC HD39.5 .A77 2025 | DDC 658.7—dc23

ASQ and ASQExcellence advance individual, organizational, and community excellence worldwide through learning, quality improvement, and knowledge exchange.

Bookstores, wholesalers, schools, libraries, businesses, and organizations: Quality Press and ASQExcellence books are available at quantity discounts for bulk purchases for business, trade, or educational uses. For more information, please contact Quality Press at 800-248-1946 or books@asq.org.

To place orders or browse the selection of all Quality Press and ASQExcellence titles, visit our website at: http://www.asq.org/quality-press.

Printed in the United States

30 29 28 27 26 25 LS 6 5 4 3 2 1

ASQ

Quality Press
600 N. Plankinton Ave.
Milwaukee, WI 53203-2914
Email: books@asq.org
Excellence Through Quality™

ASQE™
Inspire Excellence

Table of Contents

Introduction

This study guide is primarily meant to aid those taking the ASQ Certified Supplier Quality Professional (CSQP) exam. It contains 532 questions in total, drawn from sources on the exam reference list (https://www.asq.org/cert/supplier-quality/references), including *The ASQ Certified Supplier Quality Professional Handbook, Second Edition* (ASQExcellence). Section 1 provides 382 practice questions organized by the seven parts of the 2023 CSQP Body of Knowledge (BoK). Section 2 gives the reader a practice exam, comprising 150 unique questions, covering each of the seven parts of the BoK in a randomized order, that simulates the actual certification exam. Unlike other resources on the market, these questions and solutions were developed specifically to address the CSQP BoK and help those studying for the certification exam, considering the proper depth of knowledge and required levels of cognition.

Please note that all calculations for these questions were performed using a simple scientific calculator. Therefore, some answers may vary slightly if worked with a spreadsheet or other software application. Exam takers are strongly encouraged to refer to the certification FAQ (https://www.asq.org/cert/faq/faq-taking-exam) for the official ASQ calculator requirements, procure a compliant calculator, and practice using it prior to sitting for the exam. Additionally, exam takers are encouraged to purchase *The ASQ Certified Supplier Quality Professional Handbook, Second Edition* and become familiar with the content to optimize your time during the actual certification exam.

As a reminder, while the certification exam is open book, practice/sample test questions such as those in this study guide cannot be taken into ASQ certification exam rooms. It is highly recommended that you do take the Handbook and other approved references in with you to look up or verify any answers as you work the exam questions.

Your feedback and suggestions for improvement are always welcomed. Please contact me at books@asq.org, and I will do my best to clarify any questions you may have and incorporate any suggestions for improvement into future printings or editions of this study guide.

Stephanie Parker
Raleigh, North Carolina

Acknowledgments

Thank you to ASQ Quality Press, especially Lillian McAnally, Managing Editor, and Erica Barse, Associate Editor, for their expertise and technical competence, as well as their guidance through the entire publishing process. Additionally, Mark Durivage and Ed Cook are acknowledged for their work in authoring the foundational first edition of this study guide.

LIMIT OF LIABILITY/DISCLAIMER OF WARRANTY

CONTRIBUTORS' ACKNOWLEDGMENTS

The American Society for Quality and Quality Press would like to thank the Quality Press Peer Review Committee for its invaluable volunteer participation and contributions to this work. Without our volunteers' subject matter expertise, time, and passion for creating content, none of our efforts would be possible.

Quality Press Peer Review Committee Members

Scott A. Laman, Chair
Peter Pylipow, Vice Chair
Melvin Alexander
Lance Coleman
Ahmad Elshennawy
Marc Hamilton
Gary Jing
Trevor Jordan
Jane Keathley
Mary McShane-Vaughn
Jayet Moon
G.S. Sureshchandar
Tiea Theurer

Part A

Sample Questions by BoK

Section I	Supplier Strategy
Section II	Risk Management
Section III	Supplier Selection and Part Qualification
Section IV	Supplier Performance Monitoring and Improvement
Section V	Supplier Quality Management
Section VI	Relationship Management
Section VII	Business Governance, Ethics, and Compliance

This section is divided into seven parts, one for each part of the Certified Supplier Quality Professional (CSQP) Body of Knowledge (BoK). In each part of this section, there is a set of questions followed by detailed solutions.

Section I

Supplier Strategy

(59 questions)

A. SUPPLY CHAIN VISION/MISSION

Assist in the development and communication of the supply chain vision/mission statement. (Apply)

B. SUPPLIER LIFECYCLE MANAGEMENT

1. *Supplier selection.* Develop the process for supplier selection and qualification including the identification of sub-tier suppliers, using tools such as SIPOC, decision analysis, and total risk factor analysis. (Create)
2. *Performance monitoring.* Develop the supplier performance monitoring system including expected levels of performance, process reviews, performance evaluations, improvement plans, and exit strategies. (Create)
3. *Supplier classification system.* Define and develop a supplier classification system (e.g., nonapproved, conditionally approved, approved, preferred, certified, partnership, and disqualified). (Create)
4. *Partnerships and alliances.* Identify and analyze strategies for developing customer–supplier partnerships and alliances. (Analyze)

C. SUPPLY CHAIN COST ANALYSIS

1. *Cost reduction.* Identify and apply relevant inputs to prioritize cost reduction opportunities. (Analyze)
2. *Supply chain rationalization.* Interpret and analyze the optimization of a supply base to improve spending and leverage investments into supplier quality or risk reduction. (Analyze)
3. *Make/buy decisions.* Provide input on make/buy decisions by using internal and external capability analysis. Apply tools such as SWOT analysis and use historical performance to analyze requirements. (Analyze)

D. SUPPLIER AGREEMENTS OR CONTRACTS

1. *Terms and conditions.* Review and provide input for developing terms and conditions that govern supplier relationships to ensure quality considerations are addressed. (Apply)
2. *Supplier agreements.* Identify elements of supplier agreements (e.g., business and legal approach/requirements). (Understand)
3. *Quality agreements.* Analyze the elements of quality agreements/requirements (e.g., other levels of approval/review). (Analyze)
4. *Finalization controls.* Describe controls used to finalize terms and conditions that govern supplier relationships (e.g., agreements, contracts, and purchase orders). (Understand)

E. DEPLOYMENT OF STRATEGY AND EXPECTATIONS

Communicate strategy internally and communicate expectations to suppliers externally. (Apply)

QUESTIONS

1. Which of the following is/are a company's statement of its purpose and the scope of its operations?
   a. Values
   b. Vision
   c. Mission
   d. Strategy
2. Which of the following define(s) the future state of a company?
   a. Values
   b. Vision
   c. Mission
   d. Strategy

3. Which statement is an example of a vision statement?
   a. To procure the highest-quality components from the most reliable suppliers at the best cost possible
   b. To comply with all applicable standards and regulations
   c. To complete all supplier audits per the specified schedule
   d. To create a reliable supply chain through unwavering commitment to quality and cost
4. Of the following organizational concepts, which would be best associated with supporting metrics?
   a. Corporate purpose
   b. Vision statement
   c. Mission statement
   d. Objectives (strategic and tactical)
5. With the globalization of the supply chain, supplier partnerships are increasingly necessary to:
   a. comply with regulations.
   b. deliver quality products.
   c. ensure market share.
   d. provide export licenses.
6. Identification of risks in the supply chain includes all of the following activities *except*:
   a. reviewing procurement plans.
   b. analyzing assumptions and constraints.
   c. using tools such as FTA and hazard analysis.
   d. implementing prevention activities.

Use the following supplier selection table for questions 7–8:

	Criterion 1	Criterion 2	Criterion 3	Criterion 4	Criterion 5
Weight	7	3	9	5	10
Supplier A	3	3	3	3	9
Supplier B	0	9	3	3	3
Supplier C	5	5	3	5	3
Supplier D	9	3	5	3	3

7. Which supplier is ranked the lowest?
   a. A
   b. B
   c. C
   d. D
8. Which supplier should be selected?
   a. A
   b. B
   c. C
   d. D
9. Which of the following methodologies is the best for identifying stakeholders of a proposed change in the supply chain?
   a. SIPOC
   b. PDCA
   c. DMAIC
   d. DFSS
10. The SLM model is an integrated approach that considers business and quality needs of the organization. SLM is:
    a. supplier learning model.
    b. supplier life cycle management.
    c. supplier leverage model.
    d. supplier lean method.
11. The supply chain management (SCM) model focuses more on:
    a. supplier integration and costs.
    b. supplier development.
    c. supplier change management.
    d. supplier franchising opportunities.

12. During the supplier selection process for a critical component, a company is evaluating several potential suppliers. The component is essential for maintaining product quality and timely delivery. Which of the following stakeholders should be prioritized for involvement in this process?
    a. Procurement team responsible for negotiations
    b. Production team overseeing manufacturing capabilities
    c. Finance department assessing budget constraints
    d. Human resources team managing staffing needs
13. Calculate the performance index (PI) for a purchased part that costs $1000.00 and whose associated nonproductive costs amount to $100.
    a. 1.0
    b. 1.1
    c. 100
    d. 1000

Use the following table to answer questions 14–17.

	Supplier A	Supplier B	Supplier C	Supplier D
Part cost	$1750	$1650	$1835	$1800
Performance index (PI)	1.3	1.4	1.1	1.2

14. Which supplier will deliver the lowest elevated cost per part?
    a. Supplier A
    b. Supplier B
    c. Supplier C
    d. Supplier D
15. Which supplier will deliver the highest elevated cost per part?
    a. Supplier A
    b. Supplier B
    c. Supplier C
    d. Supplier D

16. What is the elevated cost for Supplier A?

    a. $2018.50

    b. $2160.00

    c. $2275.00

    d. $2310.00

17. What is the elevated cost for supplier D?

    a. $2018.50

    b. $2160.00

    c. $2275.00

    d. $2310.00

18. Which of the following tools can help in understanding the current process so that improvements can be made?

    a. FMEA

    b. Pareto chart

    c. DMAIC

    d. SIPOC

19. During which stage of the PDCA cycle would project standardization be applied to all other suppliers?

    a. Plan

    b. Do

    c. Check

    d. Act

20. During which stage of the PDCA cycle would supply chain management program development and procurement activities be started with a few selected suppliers?

    a. Plan

    b. Do

    c. Check

    d. Act

21. Which stage of the PDCA cycle would start with a spend analysis and diagnosis of the value being purchased?
    a. Plan
    b. Do
    c. Check
    d. Act
22. Which of the following indicates the need to implement an exit strategy for a particular supplier?
    a. A new SCAR has been issued.
    b. The supplier has not yet been able to complete qualification.
    c. The metrics for the supplier show long-term negative trends.
    d. The supplier has suggested changes to part finish to improve quality and decrease cost.
23. When should an exit strategy for a supplier be developed and communicated?
    a. During initial development of the contract/terms and conditions
    b. During the RFQ process
    c. When the supplier scorecard indicates long-term negative trends
    d. When the organization decides to stop buying from the supplier
24. Which of the following categories describes a supplier that meets an organization's minimum requirements?
    a. Qualified
    b. Preferred
    c. Approved
    d. Certified
25. Which of the following categories describes a supplier that has underlying performance data to establish that routine inspection of incoming materials is not needed?
    a. Preferred
    b. Qualified
    c. Certified
    d. Approved

26. All of the following should be grounds for disqualifying a supplier *except*:
    a. a long-term (6-months or more) negative trend in OTD.
    b. a SCAR issued for a new defect.
    c. repeated failure over time to follow packaging specifications.
    d. lack of responsiveness to routine communications, nonconforming material reports, and SCARs.

27. All of the following are reasons for categorizing a supplier as conditionally approved *except*:
    a. an existing supplier has recently exhibited quality and performance issues that need to be addressed.
    b. the supplier is unable to complete requalification requirements after a significant process change due to new equipment.
    c. the qualification process is not yet complete, but the outstanding items are deemed to not be critical.
    d. the requalification process is overdue due to the customer's lack of resources.

28. Which of the following describes a valid reason for bypassing the supplier qualification and approval process to purchase from a supplier?
    a. The qualification process is not complete due to a lack of resources at the organization.
    b. The supplier is already qualified to provide a similar product.
    c. The organization's customer has included a requirement to use the specific supplier in the contract with the organization.
    d. The supplier has a current, valid ISO 9001 certificate.

29. Organizations may develop alliances or partnerships with suppliers for all of the following reasons *except*:
    a. to expand core competencies.
    b. to expand into other geographic markets.
    c. to transfer risk to the supplier.
    d. to combine resources to reduce costs.

30. The costs associated with acceptance sampling are best characterized as:

    a. appraisal costs.

    b. prevention costs.

    c. internal failure costs.

    d. external failure costs.

31. The costs associated with the implementation of a companywide training initiative are best characterized as:

    a. appraisal costs.

    b. prevention costs.

    c. internal failure costs.

    d. external failure costs.

32. The acronym COPQ represents:

    a. cost of perfect quality.

    b. cost of product quality.

    c. cost of present quality.

    d. cost of poor quality.

33. There are several reasons to consider supply base optimization. Which of the following reasons would usually not be considered?

    a. Cost reduction

    b. Risk reduction

    c. Increased quantity

    d. Increased quality

34. In five years, $5000 will be available. What is the net present value (NPV) of that money, assuming an annual interest rate of 10%?

    a. $500.00

    b. $1581.14

    c. $3104.61

    d. $5000.00

35. The NPV costs to conduct a project are estimated to be $100,000. The NPV benefits or savings due to the project are estimated at $750,000. Compute the benefit-to-cost ratio.

    a. 0.13

    b. 1.3

    c. 7.5

    d. 75

36. The phrase "vital few and useful many" is applicable to the:

    a. cause-and-effect diagram.

    b. check sheet.

    c. Pareto diagram.

    d. process flow diagram.

37. Which of the following cases would be considered the least desirable in terms of the relationship of the specification band to the amount of process variation?

    a. $6\sigma < USL - LSL$

    b. $6\sigma = USL - LSL$

    c. $6\sigma > USL - LSL$

    d. $3\sigma = USL$

38. The process spread is the same as the:

    a. tolerance.

    b. quality spread.

    c. index.

    d. process variation.

39. A supplier quality engineer is performing a SWOT analysis for a new supplier offering advanced technology that could reduce production time but at a higher cost. The supplier also has limited experience in the industry. How should the engineer categorize the advanced technology and the supplier's limited experience?

    a. Advanced technology as a strength and limited experience as a threat

    b. Advanced technology as a strength and limited experience as a weakness

    c. Advanced technology as an opportunity and limited experience as a threat

    d. Advanced technology as an opportunity and limited experience as a weakness

40. Which of the following would be the preferred method/tool for identifying requirements?

    a. Purchasing document (for example, purchase order)

    b. Supply agreement

    c. (a) and (b)

    d. None of the above

41. The purchasing/procurement function has two fundamental duties. These are:

    a. to select and contract with suppliers and set terms for purchased goods and services.

    b. to ensure competitive bids and select suppliers based on cost and delivery.

    c. to source suppliers at the lowest cost and seek cost reductions going forward.

    d. to select the best supplier and negotiate a fair price.

42. Supplier performance is optimized long term through all of the following *except*:

    a. open communication.

    b. collaboration and planning.

    c. joint problem solving.

    d. ongoing price negotiations.

43. Which of the following should be used to prioritize supplier development opportunities, escalation/exit strategy, and future sourcing preferences?

    a. Pricing and schedule fulfillment

    b. Performance metrics

    c. On-time delivery scores

    d. Willingness to reduce cost year over year

44. Sourcing or purchasing can be both strategic and tactical. Which of the following is an example of a strategic function?

    a. Expediting POs in support of production

    b. Negotiating a lower price for a particular part

    c. Setting up long-term supplier partnerships

    d. Issuing a SCAR in response to failure of supplied parts in production

45. Initially, identifying and communicating roles and responsibilities with the supplier is based on:

    a. purchasing and engineering recommendations.

    b. the results of the audit.

    c. the complexity of the purchased item, and the manufacturing process required.

    d. what is needed from the supplier, based on the strength of internal core (technical) competencies.

46. Which of the following is a grouping of specific parameters that are required to ensure the ability of a product to perform as desired?

    a. Specification

    b. Contract

    c. Purchase order

    d. Statement of work

47. Which of the following is a long-term agreement for the purchase of goods and/or services from a supplier?

    a. Purchase order

    b. Statement of work

    c. Specification

    d. Contract

48. All of the following are essential elements of a contract or supplier agreement *except*:

    a. detailed description of goods/services to be provided.

    b. updated demand forecast.

    c. defined timeframe for the contract.

    d. payment terms.

49. A supplier quality engineer is drafting a new supplier agreement for a key component used in medical devices. To ensure compliance with the necessary quality standards, the agreement must clearly define the supplier's responsibilities for meeting product requirements. Which of the following is the most appropriate document to reference in the agreement for defining these requirements?

    a. Product specifications and a copy of the supplier's ISO certifications

    b. The product specifications and applicable industry standards

    c. The supplier's internal quality manual and purchase order details

    d. The product specifications, purchase orders, and shipping documents

50. All of the following are items that may be included in quality agreements with suppliers *except*:

    a. terms and conditions.

    b. process capability requirements.

    c. inspection plans.

    d. part qualification procedures.

51. A supplier quality engineer is reviewing a supplier contract for a custom component. To avoid ambiguity in product performance expectations, which is the most effective way to incorporate specifications into the terms and conditions of the agreement?

    a. Refer to the supplier's general capabilities and past performance.

    b. Include a clause that requires adherence to the company's quality manual.

    c. Attach the product drawings and specify tolerances in the contract.

    d. Include a statement that the supplier must meet *form, fit, and function* requirements.

52. A collection of documents, such as specifications, drawings, and any additional data, that define the extent of an agreement between two or more parties is referred to as _________ documents.

    a. contract

    b. engineering

    c. quality

    d. legal

53. When procuring a service, which of the following is the part of the contract where the operational aspects of the service are described?

    a. Service definition and scope

    b. Deliverables

    c. Warranties

    d. Statement of work

54. Which of the following describes a document that may be sent from the supplier to the buyer to confirm a sale?

    a. Purchase order

    b. Submittal drawing

    c. Statement of work

    d. Sales order

55. Which of the following is best described as a process to ensure that requirements are adequately identified and communicated to a supplier and that the supplier has the ability to meet those requirements?

    a. Contract review

    b. Design review

    c. Change order

    d. Supplier qualification

56. The ideal supplier management relationship is based on mutual:

    a. competition.

    b. cooperation.

    c. trust.

    d. benefit.

57. The need for a supplier typically arises in which project phase?

    a. Product development

    b. Procurement

    c. Production

    d. Prequalification

58. Once everything is agreed on between the customer and supplier, the success of the supplier deployment can be measured through:

    a. communications.

    b. monitoring.

    c. design reviews.

    d. reconciliation.

59. A supplier quality engineer is conducting feasibility analysis for a potential supplier offering a new, innovative material for use in automotive parts. The material has potential for significant cost savings but has not been widely tested under extreme temperature conditions. What is the most critical factor the engineer should evaluate to determine the feasibility of using this material?

    a. The supplier's ability to scale up production to meet mass production needs

    b. The supplier's history with similar materials in other industries

    c. The material's performance under the specific operating conditions of the automotive part

    d. The material's proven use in competitor products with similar specifications

SOLUTIONS

1. c; A company's *mission* is a statement of its purpose and its scope of operations. [I.A]

2. b; A company's *vision statement* defines the desired future state of the company. The strategy describes the actions that the company plans to take to try to achieve the vision. [I.A]

3. d; A *vision statement* is what or where the company would like to be in the future. The other answers are more task oriented and do not reflect the future state. [I.A]

4. d; Supporting objectives are usually measurable and supported by qualifiable metrics. Purpose, mission, and vision are statements. [I.A]

5. b; While complying with regulations, delivering quality products, ensuring market share, and providing export licenses all support supplier partnerships. The primary reason for supplier partnerships is to consistently deliver quality products. [I.A]

6. d; Implementation of prevention activities is a step that occurs subsequent to identification of risks. [I.B.1]

7. b; Supplier B has a weighted score of $0*7+9*2+3*9+3*5+3*10=0+18+27+15+30=90$. [I.B.1]

8. a; Suppliers A and D are tied in overall score, but the rating for Supplier A is higher than that for Supplier D for Criterion 5, which has the highest weight. [I.B.1]

9. a; Suppliers–inputs–process–outputs–customers (SIPOC) can be used to identify stakeholders. [I.B.1]

10. b; Supplier life cycle management (SLM) supports and guides the business relationship from initial supplier discovery through qualification and onboarding to ongoing maintenance and possible termination or obsolescence. [I.B.1]

11. a; Supply chain management (SCM) focuses more explicitly on the entire supply chain—including lower-tier suppliers—across the entire enterprise, and focuses on managing and minimizing total systemwide costs. [I.B.1]

12. a; Procurement ensures the selected supplier meets cost, quality, and contractual requirements. The production team does not monitor manufacturing capabilities at suppliers; this would generally be the responsibility of Procurement and Supplier Quality. [I.B.1]

13. b; $PI = \frac{\text{Purchased costs} + \text{Nonproductive costs}}{\text{Purchased costs}} = \frac{\$1000 + \$100}{\$1000} = 1.1.$ [I.B.2]

14. c; Supplier C has the lowest elevated cost per part ($2018.50).

	Supplier A	Supplier B	Supplier C	Supplier D
Part cost	$1750	$1650	$1835	$1800
Performance index (PI)	1.3	1.4	1.1	1.2
	2275	2310	2018.50	2160

[I.B.2]

15. b; Supplier B has the highest elevated cost per part ($2310).

	Supplier A	Supplier B	Supplier C	Supplier D
Part cost	$1750	$1650	$1835	$1800
Performance index (PI)	1.3	1.4	1.1	1.2
	2275	2310	2018.50	2160

[I.B.2]

16. c; The elevated cost for supplier A is $2275.

	Supplier A	Supplier B	Supplier C	Supplier D
Part cost	$1750	$1650	$1835	$1800
Performance index (PI)	1.3	1.4	1.1	1.2
	2275	2310	2018.50	2160

[I.B.2]

17. b; The elevated cost for supplier D is $2160.

	Supplier A	Supplier B	Supplier C	Supplier D
Part cost	$1750	$1650	$1835	$1800
Performance index (PI)	1.3	1.4	1.1	1.2
	2275	2310	2018.50	2160

[I.B.2]

18. d; While all of these tools may be of some use, SIPOC is a type of process map and is the best answer of the choices given. SIPOC stands for suppliers, inputs, process, outputs, and customers. The SIPOC tool is very effective in helping understand the current process so that improvements can be made. [I.B.2]

19. d; The *act* stage is where learnings from previous stages are applied anywhere else that would benefit. During the *act* stage, project standardization would be applied to all suppliers. [I.B.2]

20. b; The *do* stage is where planned actions are actually implemented. During the *do* stage, supply chain management program development and procurement activities should be started with a few selected suppliers. [I.B.2]

21. a; In the *plan* stage, a team works to understand the current state and develop an improvement plan. During the *plan* stage, a spend analysis and diagnosis of the value being purchased are evaluated. [I.B.2]

22. c; Exit strategies may be triggered by poor performance over the long-term. Issuance of a Supplier Corrective Action Request (SCAR) may not require exit if the issue is isolated and the supplier is responsive and complies with corrective action requirements. [I.B.2]

23. a; Exit strategies should be outlined up front, during negotiation and development of the initial contract/terms and conditions. The Request for Quotation (RFQ) is too early in the process, as the particular supplier hasn't been selected yet, and the other choices are too late in the process. [I.B.2]

24. c; Approved suppliers are those that meet minimum requirements. [I.B.3]

25. c; Certified suppliers have a long-term history of performance, backed up by historic data, to demonstrate that their materials do not need to be inspected when received (dock-to-stock). [I.B.3]

26. b; A SCAR *may* be grounds for disqualifying a supplier if the defect is critical and/or is a safety/liability issue, but generally a SCAR doesn't disqualify a supplier if established corrective and preventive action (CAPA) processes are followed and the supplier is responsive. The other options indicate ongoing issues that may justify disqualification. [I.B.3]

27. b; A supplier that cannot complete a requalification after a significant process change should be disqualified. A supplier with performance issues may be moved to conditional status and monitored closely. A supplier that has not completed qualification may be conditionally approved if the outstanding items are deemed to be not critical/low risk, but the qualification must eventually be completed to be fully approved. If an organization's lack of resources hinders its ability to maintain the supplier qualification process, then the process needs to be reviewed and addressed, and the lack of resources reviewed by management for action. [I.B.3]

28. c; If an organization's lack of resources hinders its ability to maintain the supplier qualification process, then the process needs to be reviewed and addressed, and the lack of resources reviewed by management for action. A supplier that is already qualified to provide a similar product could be considered to be low risk and therefore require a limited qualification, but this scenario should be accounted for in the qualification process. Suppliers with ISO 9001 certificates still need to demonstrate the technical capabilities to provide a specific product. [I.B.3]

29. c; An organization is always responsible for managing its suppliers, so it never actually rids itself of risk. [I.B.4]

30. a; *Appraisal costs* are costs associated with the inspection and appraisal processes. [I.C.1]

31. b; *Prevention costs* are the costs of all activities whose purpose is to prevent failures, including training, quality planning, and quality control activities. [I.C.1]

32. d; COPQ is the acronym used for the cost of poor quality. *COPQ* is the total of internal failure, external failure, prevention, and appraisal costs. [I.C.1]

33. c; The objectives of supply base optimization are cost reduction, risk reduction, and increased quality. [I.C.2]

34. c; In five years, \$5000 will be available. The net present value of that money, assuming an annual interest rate of 10% can be calculated by:

$$P = F(1+i)^{-n}$$

where

P = Net present value

F = Amount to be received n years from now

i = Annual interest rate expressed as a decimal

$$P = \$5000.00(1+0.10)^{-5} = \$3104.61.$$

[I.C.2]

35. c; The NPV costs to conduct a project are estimated to be \$100,000. The NPV benefits or savings due to the project are estimated at \$750,000. The benefit-to-cost ratio can be calculated by:

$$\frac{\Sigma \text{NPV of all benefits anticipated}}{\Sigma \text{NPV of all costs anticipated}} = \frac{\$750{,}000}{\$100{,}000} = 7.50.$$

[I.C.2]

36. c; The purpose of the Pareto chart is to separate the *vital few* causes from the *trivial many*. This is often reflected in what is called the 80/20 rule and helps focus attention on the more pressing issues. The 80/20 rule is a theory stating that in any situation, 20% of the inputs or activities are responsible for 80% of the outcomes or results. [I.C.3]

37. c; The least desirable relationship between the specification and the process spread is when $6\sigma > USL - LSL$. When this happens, process variation is larger than the specification range, guaranteeing that the process will produce product that is out of specification. You will want σ (process variation) to be small so that 6σ will fit well within the specification band. [I.C.3]

38. d; The process spread is simply the amount of variation present in the process. [I.C.3]

39. b; Advanced technology is a strength as it offers competitive advantages in production efficiency. Limited experience is a weakness that could affect reliability but is an internal factor. [I.C.3]

40. c; Requirements can be found on both purchase orders (POs) and supply agreements. It's important to recognize that terms and conditions can be found on a PO, sometimes on the back or sometimes on a website, but either way, they are binding. [I.D.1]

41. a; Although all four answers have more or less merit, answer (a) describes the basic purchasing function most accurately as it exists in most organizations. [I.D.1]

42. d; (a), (b), and (c) are critical to maintaining a healthy level of supplier performance. [I.B.4]

43. b; (a) and (c) are important but usually a subset of performance; without a full picture of performance, price reductions can be very dangerous and may actually increase risk. [I.C.3]

44. c; Sourcing or purchasing can be both tactical and strategic. For example, expediting POs in support of production is tactical. Setting up long-term partnerships is strategic. [I.D.1]

45. d; (a) and (b) are not accurate answers. (c) is a subset of (d)—starting to get to the root of internal competencies. (d) is most complete. [I.C.3]

46. a; This is a definition of specification. The specification is the base definition of what is being provided and is included in or referenced in the other options. [I.D.2]

47. d; This is a definition of contract (the terms "contract" and "supplier agreement" may be used interchangeably in some companies). Purchase orders are issued after the contract is established, while a statement of work and specifications are subsections or attachments that are included in or referenced by the contract. [I.D.2]

48. b; Provision of a demand forecast is a step that usually occurs after the contract has been established. [I.D.2]

49. b; Product specifications, which detail the exact characteristics of the component, and applicable industry standards ensure that the supplier is fully aware of both technical and regulatory requirements. [I.D.2]

50. a; Terms and conditions should be included in or attached to the main contract with the supplier, not the quality agreement. [I.D.3]

51. c; Referring to the supplier's past performance or capabilities is too vague and does not ensure adherence to specific product requirements. [I.D.3]

52. a; Engineering and quality documentation may be part of the contract, while "legal" is too generic a term to be a viable option here. [I.D.4]

53. d; All of the other options are subsections of the statement of work for a service. [I.D.4]

54. d; Purchase orders and statements of work flow from a buyer to a supplier, while provision of a submittal drawing (if applicable) should be done subsequent to the sales order. [I.D.4]

55. a; Design review and supplier qualification occur prior to establishment of the contract, while change orders, if needed, would occur after the order is placed and the contract review is completed. [I.D.4]

56. d; The ideal supplier management relationship is based on mutual benefit. However, trust and cooperation are also important to the relationship. [I.E]

57. a; The need for a supplier typically arises in the product development phase. The earlier the need for a supplier is determined, the easier a supplier can be selected and integrated into the design, development, and production phases. [I.E]

58. b; Once everything is agreed on, the success of the supplier deployment can be measured through monitoring of agreed-on performance metrics. [I.E]

59. c; Assessing the material's performance under the specific operating conditions, such as extreme temperatures, ensures it will function as required, making it the most critical factor in the feasibility analysis. [I.E]

Section II

Risk Management

(44 questions)

A. STRATEGY

1. *System.* Develop a risk-based approach to manage the supply base, including business continuity, contingency planning, and supply chain resilience. (Create)
2. *Product/service risk mitigation.* Develop and implement a risk mitigation plan for predicting, minimizing, and/or controlling risks. (Create)
3. *Prevention strategies.* Identify and evaluate strategies and techniques such as supply chain mapping, avoidance, detection, and mitigation used to prevent the introduction of counterfeit parts, materials, and services. (Evaluate)
4. *Supplier risk identification and categorization.* Identify supplier risks and develop categorization (e.g., organizational, business, security, and product) using tools and models, such as the Kraljic portfolio segmentation model. (Create)

B. ANALYSIS AND MITIGATION

1. *Analysis.* Identify, assess and prioritize risks to supplier quality using tools such as decision analysis (DA), failure mode and effects analysis (FMEA), fault tree analysis (FTA), and process auditing. (Evaluate)
2. *Mitigation control.* Develop and deploy controls such as inspection and test plans. Prioritize mitigation activities and sustain a risk mitigation plan appropriate to the risk of the product/service. (Create)
3. *Mitigation effectiveness.* Verify the effectiveness of the control plan and improve, if necessary, using continuous improvement methods such as plan–do–check–act (PDCA), lean, and product auditing tools. (Create)

QUESTIONS

1. A supplier quality engineer is working with the development team, which has been charged with the task of redesigning a product and reducing the overall risk to the end user. Which of the following tools would be the most appropriate for the team to use?

   a. PDCA

   b. DOE

   c. FMEA

   d. SWOT

2. When an organization is determining possible risks, they should:

   a. include all possible risks, using many risk identification tools.

   b. include only those risks that are obvious and currently known.

   c. only come up with a list of no more than three to five risks or hazards.

   d. include only risks that they have control over.

3. Residual risk is:

   a. risk that exists before a risk treatment has been implemented.

   b. risk due to a gap between what is thought to be risk and what really is risk.

   c. unknown risk that can never be identified using FMEA.

   d. risk that remains after a risk treatment has been implemented.

4. Which of the following types of plans describes strategies to allow a business to function after an unplanned event occurs?

   a. Control plan

   b. Business continuity plan

   c. Inspection plan

   d. CAPA

5. A supplier quality engineer is tasked with developing a contingency plan for a critical supplier that provides components for a medical device. The supplier is located in a region prone to natural disasters, and the company must ensure minimal disruption to production. Which combination of teams should the engineer involve in the contingency planning process to ensure all aspects of the supply chain are adequately covered?
   a. Purchasing team, quality assurance team, and finance team
   b. Logistics team, purchasing team, and quality assurance team
   c. Product design team, legal team, and finance team
   d. Engineering team, marketing team, and product design team
6. The hierarchy of risk control should follow which of the following sequences?
   a. Substitute, eliminate, engineering controls, administrative controls
   b. Substitute, eliminate, administrative controls, engineering controls
   c. Eliminate, substitute, engineering controls, administrative controls
   d. Eliminate, substitute, administrative controls, engineering controls
7. The purpose of risk evaluation is to:
   a. determine options for modifying risks.
   b. compare risk level to stated risk criteria.
   c. decide whether risk levels are acceptable.
   d. evaluate and possibly change the consequences.
8. Once the product or service has been defined and the suppliers have been selected, it is important that the manufacturer create a risk mitigation plan to minimize, monitor, and/or _________ risks.
   a. control
   b. conceive
   c. correlate
   d. connect
9. Quality risk management supports a(n) ________ approach to decision making.
   a. financial
   b. empirical
   c. procedural
   d. scientific

10. Risk assessment may be undertaken in varying degrees of depth and detail using one or several methods ranging from simple to complex. Risk assessments should consider:

    a. training, education, and experience.

    b. procedures, complexity, and training.

    c. uncertainty, resources, and complexity.

    d. residuals, resources, and training.

11. Risk assessment can be applied at all stages of ________ and is usually applied many times with different levels of detail to assist in the decisions that need to be made at each phase.

    a. product realization

    b. the product life cycle

    c. service and installation

    d. product design

12. A supplier quality engineer is responsible for detecting risks related to a new supplier that produces complex electronic components for an automotive application. Early warning signs of potential risk need to be identified to prevent production disruptions. Which of the following actions should the engineer take to detect risks early in the relationship?

    a. Implement regular supplier audits focused on financial health.

    b. Use SPC to monitor the supplier's production quality in near real time.

    c. Schedule quarterly meetings to discuss potential improvements in production.

    d. Review the supplier's ISO certification to ensure compliance with industry standards.

13. According to the Kraljic portfolio segmentation model, a product or service with low profit impact, high supply risk, and high sourcing difficulty would be considered:

    a. leverage products.

    b. strategic products.

    c. routine products.

    d. bottleneck products.

14. Strategic products deserve the most attention from purchasing managers. Options include developing long-term supply relationships, regularly analyzing and managing risks, planning for contingencies, and producing the item in-house rather than buying it, if appropriate. Strategic products are considered:

    a. low profit impact, high supply risk.

    b. low profit impact, low supply risk.

    c. high profit impact, high supply risk.

    d. high profit impact, low supply risk.

15. Purchasing approaches for routine products include using standardized products, monitoring and/or optimizing order volume, and optimizing inventory levels. Routine products are considered:

    a. low profit impact, high supply risk.

    b. low profit impact, low supply risk.

    c. high profit impact, high supply risk.

    d. high profit impact, low supply risk.

16. Useful approaches for bottleneck products include over-ordering when the item is available (lack of reliable availability is one of the most common reasons that supply is disrupted) and looking for ways to control vendors. Bottleneck products are considered:

    a. low profit impact, high supply risk.

    b. low profit impact, low supply risk.

    c. high profit impact, high supply risk.

    d. high profit impact, low supply risk.

17. Purchasing approaches to consider for leverage products include using your full purchasing power, substituting products or suppliers, and placing high-volume orders. Leverage products are considered:

    a. low profit impact, high supply risk.

    b. low profit impact, low supply risk.

    c. high profit impact, high supply risk.

    d. high profit impact, low supply risk.

18. An FMEA has a severity of 7, a probability of occurrence of 5, and a probability of detection of 3. What is the RPN?

    a. 12

    b. 15

    c. 35

    d. 105

19. A systematic approach that proactively identifies, analyzes, prioritizes, and documents potential failure modes and their respective potential causes of failures is:

    a. DFSS.

    b. FMEA.

    c. SIPOC.

    d. PDCA.

20. The likelihood that current controls will prevent a failure from reaching the customer is called:

    a. detection.

    b. occurrence.

    c. severity.

    d. exposure.

Use the following diagram to answer questions 21–23.

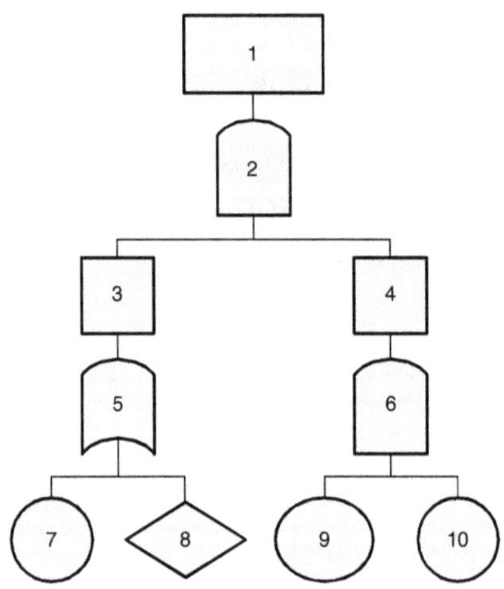

21. Which FTA symbol indicates an AND gate?
    a. 7
    b. 5
    c. 3
    d. 2
22. Which FTA symbol indicates an OR gate?
    a. 3
    b. 5
    c. 2
    d. 7
23. Which FTA symbol indicates the primary event?
    a. 1
    b. 3
    c. 5
    d. 4
24. Which of the following is best described as a systematic approach to evaluating multiple options to arrive at a best solution?
    a. Value tree analysis
    b. Decision matrix
    c. Decision analysis
    d. Influence diagram

25. Based on the following decision matrix, which option should be recommended? Higher scores are more desirable.

Solution	Criterion 1	Criterion 2	Criterion 3	Criterion 4	Criterion 5
Weight	10	8	2	5	7
Option A	1	0	1	1	5
Option B	9	9	1	9	5
Option C	5	5	5	5	1
Option D	9	9	1	9	9

a. Option A

b. Option B

c. Option C

d. Option D

26. Which of the following is described as an evaluation of an operation against specific standards to judge conformity to that standard and the effectiveness of the operation?

a. Product audit

b. Process audit

c. Service audit

d. System audit

27. A supplier quality engineer is conducting a process audit on a supplier that manufactures precision components for an aerospace application. The audit reveals that the supplier has not consistently followed its documented procedures, leading to product variability. Which of the following actions should the engineer take to identify and control potential risks?

a. Implement a more frequent audit schedule to ensure compliance.

b. Request the supplier revise their procedures to match actual practices.

c. Perform a root cause analysis to identify why procedures are not being followed.

d. Increase sampling of the supplier's products to detect defects earlier.

28. Which of the following needs to be an integral part of the risk treatment plan to ensure that the measures remain effective?
    a. Mitigation
    b. Reporting
    c. Monitoring
    d. Planning
29. Labeling, warning, identification, traceability, risk management, and recall requirements are all part of an effective:
    a. product marketing campaign.
    b. supplier quality agreement.
    c. compliance program.
    d. risk mitigation program.
30. The type of risk involved when a lot could be rejected when it does conform to the specification is known as:
    a. Ω.
    b. α.
    c. β.
    d. σ.
31. A thorough review of the inputs and outputs of an operation is essential for which of the following?
    a. Product audit
    b. Service audit
    c. System audit
    d. Process audit
32. The ability of an auditor to validate the specified audit criteria instead of just verifying them is an advantage of which of the following?
    a. Product audit
    b. GR&R
    c. Inspection planning
    d. Process audit

33. A supplier quality engineer is developing an inspection and test plan for a supplier providing electronic components. Historical data shows the following defect rates for three critical characteristics:

Characteristic	Defect Rate (PPM)	Inspection Frequency
Electrical performance	300	100% inspection
Dimensional accuracy	300	Random sampling
Surface finish	150	Random sampling

Given the defect rates, which characteristic should have its inspection frequency increased to improve quality control?

a. Electrical performance

b. Dimensional accuracy

c. Surface finish

d. Implement SPC for electrical performance instead of 100% inspection

34. Which of the following is a chart that shows, as a function of the quality of the lot being examined, the probability that a given sampling plan will result in accepting the lot?

a. OC curve

b. Sampling plan chart

c. AQL

d. AOQL

35. Which of the following is best described as the probability that a lot of AQL or better quality will be rejected?

a. Beta risk

b. AOQL

c. Power

d. Alpha risk

36. Which of the following is best described as the probability that a lot of rejectable quality will be accepted by a particular sampling plan?

a. Beta risk

b. AOQL

c. Power

d. Alpha risk

37. Analyzing the problem is associated with which phase of the PDCA cycle?
    a. Plan
    b. Do
    c. Check
    d. Act
38. Studying the results is associated with which phase of the PDCA cycle?
    a. Plan
    b. Do
    c. Check
    d. Act
39. Standardizing the solution is associated with which phase of the PDCA cycle?
    a. Plan
    b. Do
    c. Check
    d. Act
40. Identifying the opportunity is associated with which phase of the PDCA cycle?
    a. Plan
    b. Do
    c. Check
    d. Act
41. Developing the optimal solution is associated with which phase of the PDCA cycle?
    a. Plan
    b. Do
    c. Check
    d. Act

42. Implementation of which of the following techniques would be most useful to indicate when something has gone wrong so that it can be addressed and corrected?

    a. SMED

    b. Visual factory

    c. TPM

    d. Kaizen

43. Which of the following concepts refers to error-proofing or mistake-proofing?

    a. Kaizen

    b. RCA

    c. VSM

    d. Poka-yoke

44. Which of the following is described as an examination of an item to evaluate its conformance to requirements?

    a. Product audit

    b. Inspection plan

    c. CAPA

    d. Process audit

SOLUTIONS

1. c; The most appropriate tool for a team charged with the task of redesigning a product and reducing the overall risk to the end user would be Failure Mode and Effects Analysis (FMEA). [II.A.1]

2. a; The purpose of risk identification is to identify all possible risks, whether current, possible future risks, risks that are not currently under the organization's control, or risks that may occur due to the results of an accumulation of factors or steps in the process. [II.A.1]

3. d; *Residual risk* is the risk remaining even after a treatment of a known risk has been implemented. [II.A.1]

4. b; This is a definition of a business continuity plan. The others are used during regular business operations. [II.A.1]

5. b; The logistics team ensures transportation risks are addressed; the purchasing team manages supplier relations; and the quality assurance team ensures that quality standards are upheld during any transitions, making this the most balanced approach. [II.A.1]

6. c; The hierarchy of risk control should eliminate, substitute, provide engineering controls, and provide administrative controls. [II.A.2]

7. b; Risk evaluation is used to make decisions about which risks need to be addressed. [II.A.2]

8. a; Once the product or service has been defined and the suppliers have been selected, it is important that the manufacturer create a risk mitigation plan to minimize, monitor, and/or control risks. [II.A.2]

9. d; Quality risk management supports a scientific approach to decision making. [II.A.3]

10. c; Risk assessment may be undertaken in varying degrees of depth and detail using one or several methods ranging from simple to complex. Risk assessments should consider degree of uncertainty, availability of resources, and complexity. [II.A.3]

11. b; Risk assessment can be applied at all stages of the product life cycle, and is usually applied many times with different levels of detail to assist in the decisions that need to be made at each phase. Life cycle phases have different needs and require different techniques. [II.A.3]

12. b; Using Statistical Process Control (SPC) to monitor production quality allows the engineer to detect process variations or quality issues in near real-time, enabling early detection of risks. [II.A.3]

13. d; The Kraljic portfolio segmentation model classifies products with low profit impact, high supply risk, and high sourcing difficulty as bottleneck products. [II.A.4]

14. c; Strategic products are those with high profit impact and high supply risk. These products deserve the most attention from purchasing managers. Options include developing long-term supply relationships, regularly analyzing and managing risks, planning for contingencies, and producing the item in-house rather than buying, if appropriate. [II.A.4]

15. b; Routine products are those with low profit impact and low supply risk. Purchasing approaches for these items include using standardized products, monitoring and/or optimizing order volume, and optimizing inventory levels. [II.A.4]

16. a; Bottleneck products are those with low profit impact and high supply risk. Useful approaches here include over-ordering when the item is available (lack of reliable availability is one of the most common reasons that supply is disrupted) and looking for ways to control vendors. [II.A.4]

17. d; Leverage products are those with high profit impact and low supply risk. Purchasing approaches to consider here include using your full purchasing power, substituting products or suppliers, and placing high-volume orders. [II.A.4]

18. d; To calculate the Risk Priority Number (RPN) for an FMEA with a severity of 7, a probability of occurrence of 5, and a probability of detection of 3, the following formula is used:

$$RPN = S \times O \times D = 7 \times 5 \times 3 = 105$$

where

S = Severity

O = Probability of occurrence

D = Probability of detection. [II.B.1]

19. b; FMEA is a systematic approach that proactively identifies, analyzes, prioritizes, and documents potential failure modes and their respective potential causes of failures. [II.B.1]

20. a; *Detection* is the likelihood that current controls will prevent a failure from reaching the customer. [II.B.1]

21. d; Numbers 2 and 6 represent AND gates.

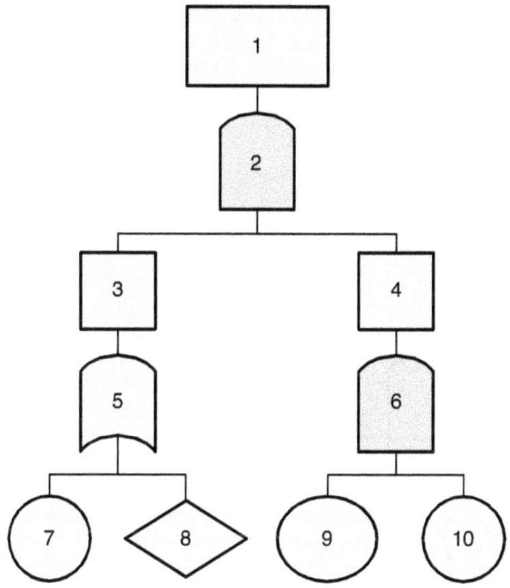

[II.B.1]

22. b; Number 5 represents an OR gate.

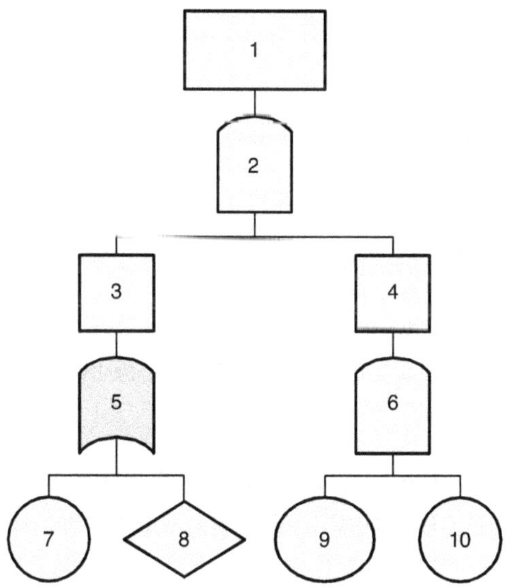

[II.B.1]

23. a; Number 1 represents a primary event.

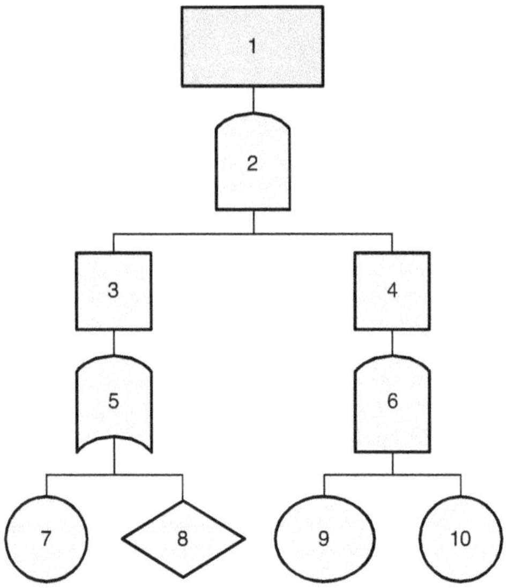

[II.B.1]

24. c; The other options are tools that may be used as part of decision analysis. [II.B.1]

25. d; Option D has the highest weighted score of the four options shown.

Option A = 10(1) + 8(0) + 2(1) + 5(1) + 7(5) = 52

Option B = 10(9) + 8(9) + 2(1) + 5(9) + 7(5) = 244

Option C = 10(5) + 8(5) + 2(5) + 5(5) + 7(1) = 132

Option D = 10(9) + 8(9) + 2(1) + 5(9) + 7(9) = 272. [II.B.1]

26. b; In this question, the key word is "operation," which indicates that a process is being audited, instead of a specific product or service or an overall management system. [II.B.1]

27. c; Conducting a root cause analysis is the best approach to understanding why procedures are not being followed and helps address the risks from process variability. [II.B.1]

28. c; Monitoring needs to be an integral part of the risk treatment plan to ensure that the measures remain effective. [II.B.2]

29. d; Labeling, warning, identification, traceability, risk management, and recall requirements are all part of an effective risk mitigation program. [II.B.2]

30. b; The type of risk involved when a lot could be rejected when it does conform to the specification is known as *alpha* (α) *risk*. [II.B.2]

Section II Solutions

31. d; A process audit includes review of a process and its associated inputs and outputs, as well as a determination of their overall validity and effectiveness. [II.B.2]

32. d; In a process audit, the auditor follows the process steps and determines the effectiveness of the process and the adequacy of the process criteria. [II.B.2]

33. b; Dimensional accuracy shares the same defect rate as electrical performance but is only randomly sampled. Increasing its inspection frequency would better capture defects and improve overall quality control. [II.B.2]

34. a; This is the definition of an Operating Characteristic (OC) curve. [II.B.2]

35. d; This is a definition of alpha risk, which is also referred to as producer's risk. [II.B.2]

36. a; This is a definition of beta risk, which is also referred to as consumer's risk. [II.B.2]

37. a; Analyzing the problem is associated with the *plan* phase of the Plan-Do-Check-Act (PDCA) cycle. You must understand the problem and its causes before taking action. [II.B.3]

38. c; Studying the results is associated with the *check* phase of the PDCA cycle. In the check phase, verify that results of implemented actions are as intended and required. [II.B.3]

39. d; Standardizing the solution is associated with the *act* phase of the PDCA cycle. Take what has been learned and apply it anywhere else that it would benefit. [II.B.3]

40. a; Identifying the opportunity is associated with the *plan* phase of the PDCA cycle. You must understand the problem and current state vs. desired state. [II.B.3]

41. a; Developing the optimal solution is associated with the *plan* phase of the PDCA cycle. Plan what to do first based on the problem identification and root cause analysis results. [II.B.3]

42. b; This is a key benefit of visual factory. The other options are other lean tools used for different purposes. [II.B.3]

43. d; This is a definition of poka-yoke. The other options are other tools used elsewhere in lean and/or continuous improvement efforts. [II.B.3]

44. a; This is a definition of a product audit. Inspection plans may be used to carry out the product audit, and a Corrective and Preventive Action (CAPA) may be initiated in response to any nonconforming condition found. [II.B.3]

Section III

Supplier Selection and Part Qualification

(74 questions)

A. PRODUCT/SERVICE REQUIREMENTS DEFINITION

1. *Internal design reviews.* Identify and apply common elements of the design review process, including roles and responsibilities of the participants. (Apply)
2. *Identifying requirements.* Identify and apply internal requirements (e.g., interrelated functional business units) for product or service in collaboration with stakeholders, including the requirements for supply chain, sub-tier suppliers, and manufacturability evaluation. (Evaluate)

B. SUPPLIER SELECTION PLANNING

1. *Supplier comparison.* Evaluate existing suppliers' and distributors' capabilities, capacities, past quality, delivery, price, lead times, and responsiveness against identified requirements. (Evaluate)
2. *Potential suppliers evaluation.* Assess potential new suppliers against identified requirements using tools such as self-assessments, audits, financial analysis, and quality function deployment. Verify third-party certification status and regulatory compliance and analyze and report on results of assessments to support the supplier selection process. (Evaluate)
3. *Supplier selection.* Evaluate and select supplier based on analysis of assessment reports and existing supplier evaluations, using decision analysis tools such as weighted decision matrices and selection matrices. (Evaluate)

C. PART, PROCESS, AND SERVICE QUALIFICATION

1. *Technical review.* Interpret and evaluate technical specification requirements and characteristics such as views, title blocks, dimensioning and tolerancing, and apply GD&T symbols as they relate to the product and process. (Evaluate)

2. *Supplier relations.* Collaborate with suppliers to define, interpret, and classify quality characteristics for the part/process/service. (Evaluate)
3. *Process and service qualification planning.* Develop a part/process/service (e.g., calibration, laboratory, software, and design) qualification plan with supplier and internal team that includes service provider audit, calibration requirements, sample size, first article inspection, measurement system analysis (MSA), process flow diagram (PFD), failure mode and effects analysis (FMEA), control plans, critical to quality (CTQ), inspection planning, capability studies, material and performance testing, appearance approval, and internal process validation. (Analyze)
4. *Part approval.* Understand production part approval process (PPAP) requirements and ensure suppliers understand the processes required to produce parts with consistent quality during an actual production run at production rates. (Understand)
5. *Validate requirements.* Collaborate with internal team to interpret the results of the executed qualification plan for the part/process/service, including reviewing Certificate of Compliance (CoC), Certificate of Analysis (CoA), and production readiness reviews (PRR). (Evaluate)

QUESTIONS

1. The APQP process is designed to be used during:
   a. the product development process.
   b. the RFQ process.
   c. the supplier qualification process.
   d. the product validation process.
2. Which of the following define(s) the phases of design, frequency of design reviews, controls on the product development process, and review and approval authorities?
   a. Design and development planning
   b. Design and development outputs
   c. Design and development changes
   d. Design and development inputs

3. Which of the following must be reviewed for adverse effects, approved, and controlled throughout the process?
   a. Design and development planning
   b. Design and development outputs
   c. Design and development changes
   d. Design and development inputs
4. The organization must ensure that which of the following satisfy/satisfies the inputs requirements?
   a. Design and development planning
   b. Design and development outputs
   c. Design and development changes
   d. Design and development inputs
5. During design and development, which of the following must consider functional and performance requirements, regulatory requirements, and safety requirements?
   a. Design and development planning
   b. Design and development outputs
   c. Design and development changes
   d. Design and development inputs
6. An often overlooked component of supplier qualification is:
   a. the audit.
   b. sub-tier suppliers.
   c. financial due diligence.
   d. the supplier questionnaire.
7. Requirements will vary based on:
   a. the cost of goods sold.
   b. the capability of the supplier to provide the purchased goods or services.
   c. the nature and complexity of the purchased goods or services.
   d. the piece part price negotiated by procurement.

8. Which of the following is a balancing process best done by internal and external stakeholders in an atmosphere of collaboration?
   a. Finalizing cost
   b. Finalizing requirements
   c. Finalizing quality agreements
   d. Finalizing preparations for an audit
9. Internal stakeholders for requirements include all of the following *except*:
   a. R&D/engineering.
   b. procurement/purchasing.
   c. operations.
   d. suppliers.
10. Procurement/purchasing function controls are applied to:
    a. prevent purchasing from going with the lowest price only.
    b. ensure the quality of the organization's procurement function.
    c. ensure outsourced product/service quality.
    d. (a) and (c).
11. Which of the following define the allowable limits of variability around the nominal value set by a product designer?
    a. Specification limits
    b. Control limits
    c. Process capability
    d. Control plans

12. A supplier quality engineer is working with a previously qualified supplier on finalizing the requirements for a new automotive part. During the process, the engineer must balance product performance, cost, and manufacturability while collaborating with various stakeholders. Which of the following should be prioritized during the finalization of the product requirements to ensure a successful launch?

    a. Ensuring that the product meets the marketing department's aesthetic goals

    b. Establishing realistic cost targets in collaboration with the procurement team

    c. Incorporating customer feedback to refine the product design before final approval

    d. Verifying that the supplier can meet the engineering team's quality standards

13. A supplier quality engineer is conducting a manufacturability evaluation for a newly designed electronic component. The following data shows key production metrics from initial trials at four different suppliers:

Supplier	Cycle Time (min)	Scrap Rate (%)	Production Cost per Unit ($)
Supplier A	12	4	8.50
Supplier B	15	2	9.00
Supplier C	10	6	8.00
Supplier D	14	3	9.50

    Based on this data, which supplier is the best choice to ensure manufacturability while maintaining cost and quality standards?

    a. Supplier A

    b. Supplier B

    c. Supplier C

    d. Supplier D

14. The process for current supplier comparison should be defined because:

    a. predefined processes ensure fairness and accuracy in comparison.

    b. predefined processes ensure consistency in application, avoiding the appearance of any impropriety.

    c. predefined processes are relational.

    d. (a) and (b).

15. Capabilities, price, capacity, quality history, and responsiveness are:
    a. great categories for a scorecard.
    b. great categories for supplier comparison.
    c. nice-to-have categories after price and delivery.
    d. great questions to ask on a desktop audit.
16. The best ranking approach uses what type of tool for analysis?
    a. Return-on-investment analysis
    b. Cost–benefit analysis
    c. Net Present Value analysis
    d. A decision matrix

For questions 17–19, refer to the following table. Higher values are more desirable.

Supplier	Price	Weight	Quality	Weight	Capacity	Weight	Totals
A	9	1	6	1	3	2	21
B	3	1	6	1	9	2	27
C	6	1	9	1	3	2	21
D	6	1	9	1	6	2	27

17. Based on the table, what's the most important category for comparison?
    a. Price
    b. Quality
    c. Capacity
    d. All are equal
18. Which supplier is most desirable?
    a. A
    b. B
    c. C
    d. D

19. Which approach does this table represent?

    a. FMEA RPN

    b. Control plan

    c. The best ranking approach

    d. Risk–benefit analysis

20. A potential supplier's quality management system (QMS) exhibits continuous monitoring and updating for necessary changes and emerging leading practices. The QMS would be considered:

    a. defined.

    b. managed.

    c. repeatable.

    d. optimized.

21. A financial audit is critical to:

    a. establish long-term economic viability and stability.

    b. establish compliance with applicable policies and regulations.

    c. quality certification.

    d. (a) and (b).

22. Which of the following tools is best used to translate customer requirements into appropriate technical requirements for each stage of product development and production?

    a. Control plan

    b. PPAP

    c. FMEA

    d. QFD

23. The following are benefits of QFD *except*:

    a. product requirements flow from customer requirements and are not misinterpreted later.

    b. the process control plan is developed and implemented.

    c. specific sales points are not lost in the translation process.

    d. misinterpretations are minimized and efficiency is increased.

24. Which of the following tools can be used to determine how well suppliers can comply with requirements for a new or revised process, product, and/or service?

  a. QFD

  b. GR&R

  c. Cpk

  d. DOE

25. During a supplier qualification audit, the supplier refuses to allow an auditor to examine the supplier's CAPA procedure, claiming that it is confidential and proprietary, but claims that the supplier holds a valid ISO 9001 certification. What should the auditor do in response?

  a. Verify that the certification is valid and that the facility/process being audited is within the certification scope.

  b. Recommend that the company discontinue doing any business with the supplier.

  c. Accept the ISO 9001 certification in lieu of audit evidence for this item.

  d. Escalate the issue to the company's management for further negotiation and action.

Choose from the following list to answer questions 26–28.

  A. A step-by-step review of inputs and outputs that relate to a specific operation, or combination of operations.

  B. A review of the implementation regulatory requirements, and adherence.

  C. A review of elements governing processes, policies, and records for quality assurance.

  D. A review of critical-to-quality (CTQ) characteristics to understand performance measures such as out-of-the-box failure.

26. A process audit is:

  a. A.

  b. B.

  c. C.

  d. D.

27. A QMS audit is:
    a. A.
    b. B.
    c. C.
    d. D.
28. A product audit is:
    a. A.
    b. B.
    c. C.
    d. D.
29. In order to make the best overall choice, supplier selection activities can all be boiled down to which of the following?
    a. Characterizing comparative risks of several factors
    b. Determining the lowest cost and highest quality
    c. Understanding the supplier's comparative capacity and lowest loss time
    d. Determining the supplier with the highest comparative capability and (potential) service levels
30. Documentation of supplier selection activities must be maintained to ensure which of the following over time?
    a. Record control
    b. The integrity of the supplier file
    c. Consistency
    d. Fairness

Use the following risk table to answer questions 31–34. Higher values are less desirable.

Supplier	QMS maturity	Weight	Value	Financial	Weight	Value	Capacity	Weight	Value	Totals
A	2	2	4	3	1	3	3	3	9	108
B	2	2	4	4	1	4	2	3	6	96
C	1	2	2	2	1	2	2	3	6	24
D	4	2	8	1	1	1	3	3	9	72

31. ________ is the most critical category.

   a. QMS maturity

   b. Capacity

   c. Financial

   d. All are equal

32. Which is the lowest-risk supplier?

   a. A

   b. B

   c. C

   d. D

33. Which is the highest-risk supplier?

   a. A

   b. B

   c. C

   d. D

34. Which supplier has the least quality risk?

   a. A and B are tied

   b. C

   c. D

   d. A

35. An engineer needs to copy an ISO A2 drawing. What sheet size (in millimeters) does the engineer need?

   a. 841×1189

   b. 594×841

   c. 420×594

   d. 297×420

36. Drawing dimensions can be linear, circular, or:
    a. rectangular.
    b. triangular.
    c. angular.
    d. singular.
37. The technical review:
    a. establishes that requirements are fully understood and accepted by suppliers.
    b. highlights technical capability of the supplier.
    c. allows for supplier input.
    d. is meaningless if the supplier objects to some or all requirements.
38. Points of discussion with the supplier in a technical review might also include:
    a. how their process can affect the finished product.
    b. future cost reductions.
    c. run rate.
    d. (a) and (c).
39. It is a best practice when concluding a technical review to:
    a. thank the supplier.
    b. review the items discussed.
    c. document the agreed-on output of the meeting.
    d. review the contract terms and conditions.
40. One very important reason to collaborate with the supplier on critical requirements is that:
    a. sometimes process control characteristics are more important to monitor than are design-critical characteristics, and the depth of knowledge at the supplier can help determine that.
    b. it's important to have a mutual agreement on quality plans, inspection, tolerances, and other requirements such as environmental controls.
    c. it helps to establish and/or deepen the relationship with the supplier, as well as obtain their buy-in.
    d. while collaboration is great, design requirements are nonnegotiable, so the relationship might need to be leveraged.

41. Collaboration on requirements adds value in that:

    a. it builds the relationship, providing a solid foundation for later in the project.

    b. it can provide the supplier with knowledge of how the customer will use their product.

    c. it provides a healthy give-and-take on both parties' behalf.

    d. it helps the supplier feel ownership and encourages buy-in.

42. Which of the following is represented by the symbol ▱?

    a. Straightness

    b. Parallelism

    c. Perpendicularity

    d. Flatness

43. In GD&T notation, the symbol ø represents which of the following?

    a. Diameter

    b. Circularity

    c. Concentricity

    d. Cylindricity

44. Which of the following is represented by the symbol ↧?

    a. Depth

    b. Perpendicularity

    c. Symmetry

    d. True Position

Use the following figure to answer question 45.

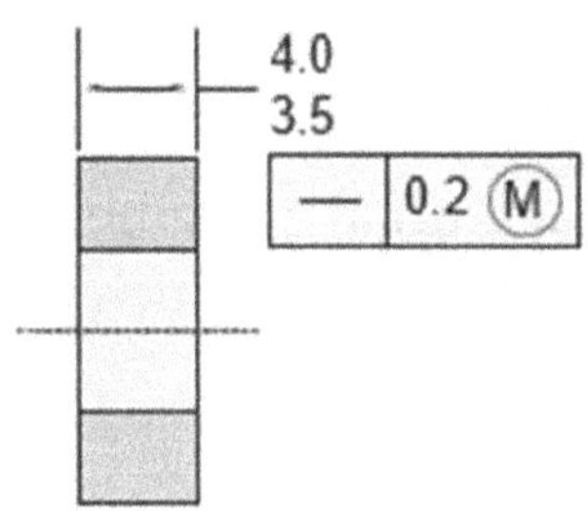

45. What is the bonus tolerance if the feature is 3.7?

a. 0.0

b. 0.2

c. 0.3

d. 0.5

Use the following figure to answer question 46.

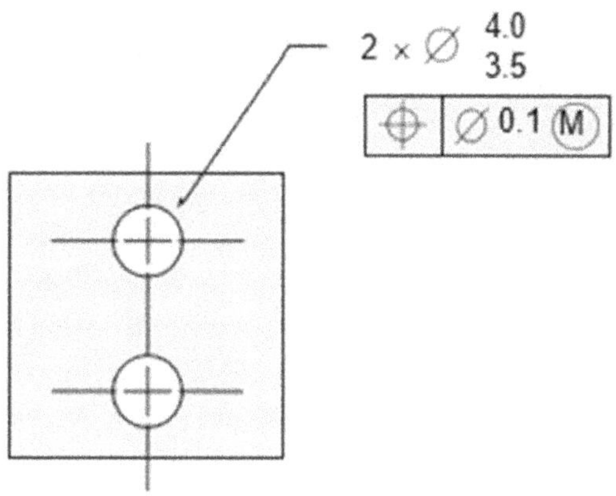

46. What is the bonus tolerance if the feature is 3.7?

a. 0.0

b. 0.2

c. 0.3

d. 0.5

47. Which of the following refers to the activities necessary to control measuring systems to ensure that they are functioning within prescribed accuracy objectives?

a. GR&R

b. Capability study

c. Calibration

d. Inspection plan

48. A supplier notifies your company that they have found that a device used to measure some of your product is out of calibration. What do you do next?

  a. Review data from Certificates of Analysis to determine next steps.
  b. Initiate a SCAR.
  c. Identify and segregate affected material from the supplier.
  d. No further action is required.

49. Which of the following is a technique for measuring the properties and features of an initial product sample against the dimensions and specifications on an engineering drawing?

  a. GR&R
  b. FAI
  c. Cpk
  d. GD&T

50. A PFD does which of the following?

  a. Controls the process
  b. Diagrams a lean material system using the principle of flow
  c. Maps the process steps and sequence of operations in a process
  d. Displays where product goes in the value stream

51. A PFMEA is a _______ tool.

  a. risk management
  b. quality management
  c. sales forecasting
  d. quality engineering

52. What is the primary relationship between the PFD and the control plan?

  a. The control plan must be created first, as it plans the control process.
  b. There is no relationship; they are independent.
  c. They are both components of PPAP.
  d. The control plan addresses each operating step laid out in the PFD.

53. The PPAP:
    a. has no value outside of the automotive industry.
    b. establishes that components and subassemblies meet requirements.
    c. is too difficult for most suppliers.
    d. includes the elements of quoting, costing, and demand planning.
54. GR&R studies are important because:
    a. they establish measurement system error and ensure that the data we collect are accurate and precise.
    b. they prove through objective evidence whether an inspector is proficient.
    c. they qualify inspectors to complete PPAP.
    d. they highlight the differences in technique between inspectors.
55. CTQ characteristics:
    a. are critical to quality.
    b. highlight technical capability of the supplier.
    c. allow for supplier input.
    d. (a) and (c).
56. Which of the following analyzes variation in measurements attributable to gages and operators?
    a. Cpk
    b. QFD
    c. MSA
    d. FAI
57. Which of the following is measurement variation attributed to the gage used?
    a. Reproducibility
    b. Accuracy
    c. Repeatability
    d. Precision

58. Which of the following is measurement variation attributed to the gage operator?

a. Reproducibility

b. Repeatability

c. Accuracy

d. Precision

59. Which of the following is the foundation of PFMEA, HACCP, and control plans?

a. VSM

b. PFD

c. QFD

d. MSA

60. For an FMEA using an Action Priority Table instead of RPN, which of the following ratings is emphasized the most when prioritizing actions?

a. Frequency

b. Occurrence

c. Severity

d. Risk

61. Which of the following contains information regarding process characteristics to be evaluated, requirements and specifications for those characteristics, the methods for evaluating process output, and actions to take if a nonconforming result is detected?

a. Control plan

b. SIPOC diagram

c. FMEA

d. Process value chain diagram

62. A company inspecting lots received from a supplier would switch from Reduced to Normal inspection level for all of the following situations except:

a. A lot is not accepted.

b. Five consecutive lots are accepted.

c. A lot is accepted but with a nonconformity quantity between Ac and Re on the AQL table.

d. Production is irregular or delayed.

63. When should a company switch from Normal to Tightened inspection levels?

    a. Five consecutive lots have been deemed acceptable at original inspection.

    b. The preceding 10 lots have not been rejected on original inspection.

    c. Two of five consecutive lots have been rejected at original inspection.

    d. Any lot is rejected at original inspection.

64. All of the following are acceptable inputs to determining an appropriate AQL level *except*:

    a. statistical convenience.

    b. needs of the customer.

    c. needs of the producer.

    d. the break-even point.

65. The technical review:

    a. is not part of PPAP.

    b. highlights technical capability to the supplier.

    c. allows for supplier input.

    d. is a one-way conversation.

66. PPAP is required in all of the following situations *except*:

    a. new parts.

    b. pricing changes.

    c. changes in part processing.

    d. sub-tier supplier or materials change.

67. The specific phrase, "establishing through objective evidence that a process consistently produces a result or product meeting its predetermined requirements" refers to:

    a. process validation.

    b. a qualified process.

    c. PPAP.

    d. process capability.

68. Process validation consists of:

    a. running at rate, RFQ, and process optimization.

    b. control plans, PSW, and process map.

    c. material performance tests, biocompatibility, and sterilization.

    d. IQ, OQ, and PQ.

69. A supplier quality engineer is tasked with applying CTQ characteristics for a new medical device component. The CTQs include stringent dimensional accuracy and material composition requirements. What is the best approach to ensure that the supplier consistently meets these CTQs during production?

    a. Increase the frequency of supplier audits to monitor compliance with CTQ requirements.

    b. Implement SPC for the CTQ dimensions and conduct regular material testing.

    c. Focus on reducing production costs while maintaining general product quality standards.

    d. Rely on final product inspections to identify any deviations from CTQ specifications.

70. As a rule, the internal (customer) cross-functional team working on part/process approval should consist of:

    a. supplier, engineering, and QA.

    b. QA, engineering, and operations.

    c. sales, purchasing, and operations.

    d. supplier, QA, and purchasing.

71. Collaboration with the internal team becomes especially important when:

    a. deviations from the qualification plan occur.

    b. the supplier raises prices.

    c. the supplier produces a quality part but at less than the rate required to meet demand.

    d. (a) and (c).

72. Critical elements of PPAP to review and approve are:
    a. control plan, MSA, PFMEA, PFD, and capability study.
    b. IQ, OQ, and PQ.
    c. PPV, run at rate, and tooling cost.
    d. CTQs, safety program, and employees training program.

73. When reviewing capability studies, it's important that we know that:
    a. the inspectors are adequately trained, so we can rely on the results.
    b. the process is in a state of statistical control.
    c. the resulting frequency distribution is normally distributed.
    d. (b) and (c).

74. A supplier quality engineer is participating in a production readiness review (PRR) for a new automotive component. The review is intended to ensure that the supplier is ready for full-scale production. Which of the following is the most critical factor to evaluate during the PRR?
    a. The supplier's ability to meet delivery schedules with existing resources
    b. The supplier's past performance in delivering similar components
    c. The supplier's readiness to consistently meet CTQ requirements
    d. The supplier's willingness to renegotiate pricing before production

SOLUTIONS

1. a; Advanced Product Quality Planning (APQP) is not related to Request for Quotation (RFQ), but validation might be a part of it. Supplier qualification might be a part of the development process. But (a) is the right answer; it is a quality process to be used during the design and development process. [III.A.1]
2. a; Planning determines the scope and deliverables of the design process. [III.A.1]
3. c; Changes must be controlled and reviewed for impact and unintended or adverse effects. [III.A.1]
4. b; Outputs must satisfy inputs requirements. It's how we know we've succeeded. [III.A.1]
5. d; Inputs define the product and must meet a whole host of requirements as well as satisfy the design intent. [III.A.1]
6. b; Sub-tier suppliers can be critical to quality. Understand the level and nature of outsourcing by your suppliers. Financial assessment is rarely forgotten and (a) and (d) are very similar. [III.A.2]
7. c; Requirements are sacrosanct and are based on design outputs, as determined by the nature and complexity of your purchase. Requirements should never be altered because of price or supplier capability. [III.A.2]
8. b; Finalizing requirements is the only process listed that involves all the internal and external stakeholders. It is intended to balance stakeholder needs in order to meet goals for quality, cost, and delivery. [III.A.2]
9. d; The groups in choices A, B, and C need to be considered internal stakeholders along with supplier quality and any local purchasing or materials planners. Suppliers are *external* stakeholders. [III.A.2]
10. d; Both ensure the procurement process and the outsourced product/service quality. [III.A.2]
11. a; This is the definition of specification limits. Control plans contain references to the specification limits, but they are usually actually defined elsewhere. [III.A.2]
12. b; Collaboration with procurement is essential to ensure that the final product meets cost goals while maintaining performance, quality, and delivery targets, crucial for balancing competing objectives during product development. The previous successful qualification of the supplier should have already demonstrated that the supplier can meet standards. [III.A.2]
13. b; Supplier B provides the lowest scrap rate, ensuring higher quality and less waste. Despite a slightly higher cost per unit and cycle time, the reduced defects make them the best option for long-term manufacturability. [III.A.2]

14. d; (a) and (b) are both very good reasons to use a defined process to come to the best and most defensible decision. [III.B.1]

15. b; These are categories for comparison when deciding who to give additional business to. Capacity is not usually a scorecard (monitoring) category. Quality history is not usually a category on a questionnaire or desktop audit either, as that information does not come from the supplier, but the customer (you). [III.B.1]

16. d; This approach uses a simple table with weighted values and comparison categories. [III.B.1]

17. c; Since capacity is weighted higher, it would be the most critical category. [III.B.1]

18. b; Since there is a numerical tie, and capacity is weighted higher, supplier B would be the choice. [III.B.1]

19. c; This best ranking approach uses a table to compare and tabulate numerical rankings based on the most important categories. [III.B.1]

20. d; A quality management system (QMS) that exhibits continuous monitoring and updating for necessary changes and emerging leading practices would be considered to be optimized. [III.B.2]

21. d; Both compliance (especially with anticorruption laws) and financial viability are important reasons to perform due diligence. [III.B.2]

22. d; Quality function deployment (QFD) translates customer requirements into actual technical requirements. Control plans and Failure Mode and Effects Analyses (FMEAs) may be used later when developing production processes; Production Part Approval Process (PPAP) is used to qualify a supplier. [III.B.2]

23. b; The process control plan would be developed in a subsequent step, after the QFD is completed. [III.B.2]

24. a; The key phrase is "new or revised," which implies design activity. Of the options given, QFD is best used during design. [III.B.2]

25. a; ISO 9001 requires that an organization has a process for managing corrective actions and maintaining documented evidence. If the auditor verifies that the supplier's certificate is valid and that the scope covers the site/process that is being audited, then this can be considered evidence that the Corrective and Preventive Action (CAPA) process exists and has been judged to be effective. The auditor's management may wish to discuss management of proprietary information further and/or complete a Non-Disclosure Agreement (NDA) with the supplier after the audit is complete, but that action would be subsequent to the audit. [III.B.2]

26. a; A focus on inputs, outputs, and steps is an indication that a process is being audited. [III.B.2]

27. c; A focus on quality assurance and policies is an indication that a QMS is being audited. [III.B.2]

28. d; Critical-to-quality (CTQ) and out-of-the-box should lead the reader to understand that a product audit is underway. [III.B.2]

29. a; "Overall" is the operative word, so it is about risk over many categories. [III.B.3]

30. c; To ensure fairness, consistency is the key. While these activities are records for the supplier file, that's not the key concept. [III.B.3]

31. b; Capacity is weighted at a 3, the highest of all the categories. [III.B.3]

32. c; (c) is the lowest at 24. [III.B.3]

33. a; (a) is the highest by far; the risk value is 108. The next closest is (b) at 96. [III.B.3]

34. c; (c) is the lowest risk at 1. (a) and (b) are tied at 2. [III.B.3]

35. c; An ISO A2 drawing sheet is 420 mm × 594 mm. [III.C.1]

36. c; Drawing dimensions can be linear, circular, or angular. [III.C.1]

37. a; The technical review is critical to open dialogue with suppliers, obtain feedback and commitment to requirements, ensure supplier capability, and get a process reality check. [III.C.1]

38. d; Future cost reductions are not an element of technical review. [III.C.1]

39. c; It is *always* a best practice to document results of a supplier meeting, but more so as part of technical review. [III.C.1]

40. a; All four answers are correct to some degree, in some context, but once again the real value of collaboration is tapping into the profound process knowledge of your supplier. [III.C.2]

41. b; (a) is a great answer, but (b) provides more value. (c) and (d) are true but not specifically value-added regarding requirements. [III.C.2]

42. d; In Geometric Dimensioning and Tolerancing (GD&T), a flatness tolerance defines a zone between two parallel planes within which a surface must lie. [III.C.2]

43. a; In GD&T, the diameter symbol indicates that the size of a circular feature is being dimensioned using the diameter of that feature. [III.C.2]

44. a; The depth symbol is used to indicate a measurement from the bottom of a feature to the outer surface of a part. [III.C.2]

45. c; Bonus tolerance = Absolute difference between maximum material condition (MMC) and actual condition. BT = 4.0 – 3.7 = 0.3. [III.C.2]

46. b; Bonus tolerance = Absolute difference between MMC and actual condition. BT = 3.5 − 3.7 = 0.2. [III.C.2]

47. c; Each option refers to parts of measuring and inspection activities, but calibration is fundamental to all of the others. [III.C.3]

48. c; While the supplier works to determine the scope of material that is affected, any suspect material from the supplier should be immediately identified and segregated to prevent its use before disposition. As the supplier is working through their own processes for control of nonconforming material and notification of customers, a Supplier Corrective Action Request (SCAR) may not be necessary, although you may choose to request a copy of the supplier's internal corrective action report and findings. [III.C.3]

49. b; This is the definition of first article inspection (FAI). [III.C.3]

50. c; A Process Flow Diagram (PFD) is a diagram of steps in the process and can be used for any process. [III.C.3]

51. a; A Process Failure Mode and Effects Analysis (PFMEA) is a tool characterizing risks and their mitigations following the PFD. [III.C.3]

52. d; A control plan addresses unmitigated risks from the PFMEA, as delineated step-by-step from the PFD. [III.C.3]

53. b; PPAP is a series of quality tools designed to show that requirements are met via Measurement System Analysis (MSA), capability studies, Statistical Process Control (SPC), and PFMEA. [III.C.3]

54. a; Gauge Repeatability and Reproducibility (GR&R) is critical in establishing the validity of future data because it determines error in the entire measurement system: gages, staging and fixtures, and operators (inspectors). [III.C.3]

55. d; CTQs are the critical-to-quality characteristics, and it is wise to allow supplier input on this determination, as process parameters can also be CTQ. [III.C.3]

56. c; This is the main purpose of MSA. Quality Function Deployment (QFD) identifies specifications and requirements that may lead to identification of required gaging; gages must be verified as appropriate for use using MSA before conducting FAI or Cpk analysis. [III.C.3]

57. c; Repeatability is the ability of the same operator to get the same result using the same gage on the same piece being measured. [III.C.3]

58. c; Reproducibility is the ability of different operators to get the same result using the same gage on the same piece being measured. [III.C.3]

59. b; The PFD is used as the basis for PFMEA, Hazard Analysis and Critical Control Points (HACCP), control plans, and Value Stream Mapping (VSM). [III.C.3]

60. c; When using an Action Priority Table, the Severity rating is emphasized over the others as it rates the level of direct impact to the user. [III.C.3]

61. a; This is the definition of a control plan. Suppliers-Inputs-Process-Outputs-Customers (SIPOC) diagrams and/or value chain diagrams may be used to help prepare a control plan, while FMEAs are used more in risk analysis and mitigation. [III.C.3]

62. b; The other choices are included in criteria for switching from Reduced to Normal inspection levels in the ANSI Z1.4-2008 standard, while (b) is a criterion for switching from Tightened to Normal inspection levels. [III.C.3]

63. c; (a) is a criterion for switching between Tightened to Normal inspection levels, (b) is a criterion for switching between Normal and Reduced inspection levels, and (d) is a criterion for switching between Reduced and Normal inspection levels. [III.C.3]

64. a; Statistical convenience must *never* be a reason for determining an appropriate Acceptance Quality Limit (AQL) level. The break-even point is the cost to inspect one piece divided by the damage done by one defective; this value may help in determining an appropriate AQL. [III.C.3]

65. c; Technical review is a critical component of PPAP and is intended as a two-way dialogue with the supplier. [III.C.4]

66. b; (a), (c), and (d) are all conditions where PPAP should be performed. Pricing alone is not. [III.C.4]

67. a; Process validation includes capability studies and certainly is used to qualify processes. PPAP is an automotive tool focused on part approval. Validation is a biomedical/pharmaceutical tool that focuses on the process to ensure part quality. [III.C.4]

68. d; Installation Qualification (IQ), Operational Qualification (OQ), and Performance Qualification (PQ) are specific to process validation for medical device/pharma products. [III.C.4]

69. b; SPC provides ongoing control of critical dimensions in near real time; regular material testing ensures that the CTQs are met consistently. [III.C.4]

70. b; Since part and process qualification is technical, QA, engineering, and operations are the right team, at a minimum, to evaluate the supplier's work. [III.C.5]

71. d; Deviations and failure to do *run at rate* are critical and must be discussed cross-functionally. [III.C.5]

72. a; IQ, OQ, and PQ are never PPAP elements. Purchase Price Variance (PPV) and tooling cost are not PPAP elements, but purchasing data. CTQs and safety are great audit elements. [III.C.5]

73. d; Capability indices (Cpk, Cp) are only valid under the assumptions of control and normality. GR&R shows measurement system quality (including inspector repeatability). [III.C.5]

74. c; Ensuring the supplier is fully prepared to meet CTQ requirements is critical to the success of full-scale production. [III.C.5]

Section IV

Supplier Performance Monitoring and Improvement

(66 questions)

A. SUPPLIER PERFORMANCE MONITORING

1. *Supplier metrics.* Define, implement, and monitor supplier performance metrics such as quality, delivery (e.g., on-time delivery [OTD] and on-time in full delivery [OTIF]), cost, and responsiveness. (Evaluate)
2. *Supplier performance.* Analyze supplier performance data (e.g., warranty analysis/field returns and defect rates) and develop periodic reports (e.g., scorecard and dashboards). (Analyze)
3. *Supplier process performance.* Define and implement lean principles and applications such as 5S, kaizen, value stream mapping, supplier process capabilities and controls, 8 wastes, single minute exchange of dies (SMED), kanban, muda, standardized work, takt time, and error-proofing to reduce waste and increase performance. (Evaluate)

B. ASSESS NONCONFORMING PRODUCT/PROCESS/SERVICE

Segregate, control, and evaluate nonconforming materials to determine whether a material review board (MRB) requires disposition. Conduct risk assessments to prevent future discrepancies. (Evaluate)

C. SUPPLIER CORRECTIVE AND PREVENTIVE ACTION (CAPA)

1. *Root cause analysis tools and methods.* Evaluate the root cause analysis of a problem using tools such as cause and effect diagrams, Pareto analysis, 5 Why's, fault tree analysis, design of experiments (DOE), brainstorming, check sheets, measurement system analysis (MSA), production records, and review of process flow. (Evaluate)
2. *Collaboration with supplier.* Evaluate and implement supplier corrective/preventive action and review its effectiveness and robustness with supplier. Understand the process of updating failure mode and effects analysis (FMEA) and process control plan, and understand statistical process control (SPC), 8D, and product/process design change. (Evaluate)

QUESTIONS

1. Incoming acceptance can be done through either lot or part acceptance. Acme Manufacturing, a supplier of key components, shipped 27 lots in Q1, of which one was rejected. What is the percentage of lot acceptance for Q1?

   a. 3.70

   b. 3.85

   c. 96.30

   d. 100.00

2. Incoming acceptance can be done through either lot or part acceptance. Acme Manufacturing, a supplier of key components, shipped 27 lots in Q1, of which one was rejected. What is the percentage of lots rejected for Q1?

   a. 3.70

   b. 3.85

   c. 96.30

   d. 100.00

3. In-process acceptance is calculated by part acceptance rates. Acme Manufacturing, a supplier of key components, provided the following information for Q4: 768 parts were manufactured, of which 17 parts were found to be incorrect and had to be discarded. What is the in-process acceptance percentage rate for Q4?

   a. 2.21

   b. 45.18

   c. 97.79

   d. 100.00

Use the following information to answer questions 4–5.

Number of orders total = 40

Number of orders delivered on time = 38

Number of orders delivered in full on time = 35

4. What is the OTD percentage?
   a. 76.0%
   b. 87.5%
   c. 95.0%
   d. 97.5%

5. What is the OTIF percentage?
   a. 76.0%
   b. 87.5%
   c. 95.0%
   d. 97.5%

6. Consider the following formula:

$$\frac{\text{Total orders completed in full on time}}{\text{Total orders completed}}$$

   Which of these metrics is calculated using this formula?
   a. OTIF
   b. OTD
   c. FPY
   d. DPMO

7. A small manufacturing company uses a risk-based SCAR decision matrix. To effectively use the matrix, what is the proper process flow?
   a. Identify the severity-frequency rating, determine the frequency, identify the risk severity, take appropriate action
   b. Identify the severity-frequency rating, determine the frequency, identify the risk severity, take appropriate action
   c. Identify the risk severity, identify the severity-frequency rating, determine the frequency, take appropriate action
   d. Identify the risk severity, determine the frequency, identify the severity-frequency rating, take appropriate action

8. What does the acronym SCAR mean?
   a. Single corrective action request
   b. Supplier corrective action request
   c. Single corrective action requirement
   d. Supplier corrective action requirement
9. Which tool is best used for formally communicating quality issues with a supplier?
   a. DOE
   b. DMAIC
   c. SCAR
   d. PDCA

Use the following information to answer questions 10–12.

A supplier quality engineer is completing the quarterly scorecard for a high-risk supplier using the following criteria.

Product quality:

$$\frac{\text{Total number of line items} - \text{Number of SCARs}}{\text{Total number of line items}} \times 60$$

On-time delivery:

$$\frac{\text{Number of on-time deliveries}}{\text{Total number of line items}} \times 40$$

Scorecard rating = Product quality + On-time delivery

Total line items for Q3: 25 items

Total SCARs: 3

Number of on-time deliveries: 20

10. The product quality score is:
    a. 32.0.
    b. 40.0.
    c. 52.8.
    d. 60.0.

11. The on-time delivery score is:

    a. 32.0.

    b. 40.0.

    c. 52.8.

    d. 60.0.

12. The supplier Q3 scorecard rating is:

    a. 40.0.

    b. 60.0.

    c. 84.8.

    d. 100.0.

13. Which of the following utilizes supplier performance and capability information to identify suppliers whose materials can be received directly without verification?

    a. Just-in-time

    b. Source inspection

    c. Supplier performance index

    d. Ship-to-stock

14. Lean manufacturing principles are the foundation of any successful manufacturing company. The implementation of these principles can be done in any order as long as the implementation is effective and sustainable. Which of the following is least likely to be used for a lean initiative?

    a. 5S

    b. Kanban

    c. Standard work

    d. Design of experiments

15. Single-minute exchange of die (SMED) generally has a target time of _____ minute(s).

    a. 1

    b. 10

    c. 30

    d. 60

16. When developing work instructions, best practice includes the use of:
    a. figures.
    b. tables.
    c. narrative.
    d. photographs.
17. Which of the following is considered by many experts to be the sixth S in a 5S program?
    a. Safety
    b. Sort
    c. Shine
    d. Sustain
18. A supplier is using a *pull* system for managing inventory. Inventory management pull systems are also known as:
    a. Hoshin
    b. Kaizen
    c. Kanban
    d. SMED
19. Costs associated with inventory are generally considered to be approximately ____ of the actual total dollar value at the component or piece part level.
    a. 10–20%
    b. 30–40%
    c. 50–60%
    d. 60–70%
20. When developing standard work, the scope of work should be outlined or documented on a:
    a. check sheet.
    b. kanban card.
    c. flowchart.
    d. control plan.

21. Procedures and work instructions require proper:

    a. credentials.

    b. formatting.

    c. pagination.

    d. revision control.

22. For general manufacturing, average inventory turns range from:

    a. one to three.

    b. three to five.

    c. five to seven.

    d. seven to nine.

23. Kanban uses the _____ to determine the right amount of inventory. Once that quantity is consumed, the signal is sent to the supplier to replenish the quantity required.

    a. standard work rate

    b. production consumption rate

    c. postproduction sorting rate

    d. delivery requirements

24. In the 5S methodology, the Japanese word for cleaning the work area is:

    a. seiri.

    b. seiton.

    c. seiso.

    d. seiketsu.

25. A customer order for 1000 quarters is due in five days. ABC runs one eight-hour shift per day. What lean tool will help with reducing the impact of changeovers on the process?

    a. Poka-yoke

    b. SMED

    c. 5S

    d. Standard work

26. Which quality guru developed the kanban concept?

    a. Genichi Taguchi

    b. Masaaki Imai

    c. Taiichi Ohno

    d. W. Edwards Deming

27. Producing a punch tool that reduces costs and cycle time by creating all part features at the same time the diameter is being punched is an example of:

    a. standard work.

    b. a kaizen event.

    c. poka-yoke.

    d. takt time.

28. A customer order for 1000 quarters is due in five days. ABC runs one eight-hour shift per day. What is the takt time (in minutes per unit)?

    a. .417

    b. 2.4

    c. 25

    d. 625

29. Which of the following is another name for poka-yoke?

    a. Waste

    b. Error-proofing

    c. Orderliness

    d. Pull system

30. Which of the following wastes occurs when more product is made than is needed by the next process?

    a. Overproduction

    b. Transportation

    c. Motion

    d. Downtime

31. The principle that each process should be done the same way, every time, by every employee is called:

    a. kanban.

    b. 5S.

    c. standardized work.

    d. pull system.

32. Which of the following tools is typically used first in a lean system?

    a. 5S

    b. Kanban

    c. Value stream map

    d. Visual control

33. In the 5S methodology, the Japanese word for sorting is:

    a. seiri.

    b. seiton.

    c. seiso.

    d. seiketsu.

34. What purpose does the MRB of a company serve?

    a. Overseeing the traceability of a product

    b. Developing sampling plans for product materials

    c. Identifying the source of all product components

    d. Determining disposition for nonconforming components

35. Keeping nonconforming products from being mixed with good/conforming products within a manufacturing plant is known as:

    a. compliance.

    b. material segregation.

    c. material review.

    d. defect classification.

36. Customers have complained about the battery life of a new line of cell phones released. How would the company classify this type of product defect?

    a. Minor defect

    b. Major defect

    c. Serious defect

    d. Critical defect

Use the following information to answer questions 37–41.

A lot containing 100 printed circuit boards (PCBs) was received and sent to incoming inspection. The inspection plan called for 15 randomly selected boards to be tested. During incoming inspection, the 15 randomly selected PCBs failed the "burn-in" test. The 15 PCBs were immediately labeled, placed on electronic hold, and moved to the quarantine storage area. A nonconformance report was initiated and placed on the agenda for the material review board.

37. What is the first action the MRB should perform?

    a. Issue a SCAR to the supplier.

    b. Ensure that the remaining 85 PCBs are inspected.

    c. Issue a rework order.

    d. Ensure that the remaining 85 PCBs are quarantined.

38. The PCBs are critical to the product being manufactured. During the MRB meeting, the incoming inspection manager reported that the last lot of this particular PCB was not inspected due to the skip-lot sampling scheme. What action should the MRB perform?

    a. Issue a SCAR to the supplier.

    b. Release the remaining 85 PCBs to production.

    c. Place all PCBs from the previous lot on hold.

    d. Issue a rework order.

39. The production manager is worried that the month's production quota will not be met and wants the MRB to release the PCBs from both lots to production and to ship the equipment that was assembled using the PCBs from the skip-lot. What is the most reasonable action for the MRB?

    a. Issue a SCAR to the supplier.

    b. Nothing; the MRB should wait for the supplier's response.

    c. Perform a retest based on an AQL sampling on the equipment that was assembled using the PCBs from the skip-lot and release.

    d. Perform a retest of all of the equipment that was assembled using the PCBs from the skip-lot and release.

40. The supplier of the PCBs informs the MRB that a new manufacturer of a component was utilized beginning with the lot of 100 PCBs that failed incoming inspection. What is the most reasonable action for the MRB?

    a. Issue a SCAR to the supplier.

    b. Nothing; the MRB should wait for the supplier's response.

    c. Perform a retest based on an AQL sampling of the equipment that was assembled using the PCBs from the skip-lot and release.

    d. Perform a retest of all of the equipment that was assembled using the PCBs from the skip-lot and release.

41. Which of the following would be an acceptable response from the supplier?

    a. Provide change notification.

    b. Have the supplier audit their supplier.

    c. Have the supplier change suppliers.

    d. Nothing; this was a one-time issue.

42. A visual identification of many potential causes of a problem is referred to as a:

    a. process flowchart.

    b. cause-and-effect diagram.

    c. decision tree.

    d. check sheet.

43. A process improvement team has completed a PFMEA. Which of the following tools would best be suited to help the team focus their efforts based on the PFMEA?

    a. Pareto analysis

    b. Force-field analysis

    c. SWOT analysis

    d. Statistical analysis

44. A team was put together to analyze potential actions that could lead to process failures. The goal is for the team to help determine the risk associated with process failures and to prevent future failures from occurring. Which quality tool would help the team visualize this scenario?

    a. Fault tree

    b. Affinity diagram

    c. Matrix diagram

    d. Process decision program chart

A three-factor full factorial experiment was conducted. Each factor has two levels. Use the following figure to answer questions 45–46.

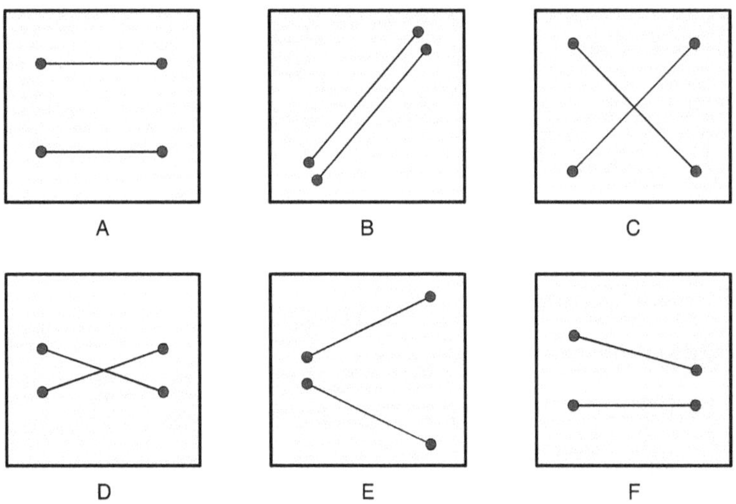

45. Which of the following plots do not indicate the presence of an interaction?

    a. A and B

    b. C and D

    c. E and F

    d. None of these

46. Which of the following plots indicate the most significant interactions?

    a. A and B

    b. B and F

    c. A and F

    d. C and D

47. A technique used by supplier quality engineers for the generation of ideas—where ideas are written down without any discussion—is:

    a. brainstorming.

    b. affinity analysis.

    c. nominal group technique.

    d. configuration management.

48. The tool that can be used to organize the ideas developed during a brainstorming session into larger categories is the:

    a. interrelationship diagram.

    b. Gantt chart.

    c. affinity diagram.

    d. work breakdown structure.

49. If the bias of a measuring system increases across the measurement range, the measuring device is said to have poor:

    a. stability.

    b. linearity.

    c. bias.

    d. reproducibility.

Use the following figure to answer questions 50–51.

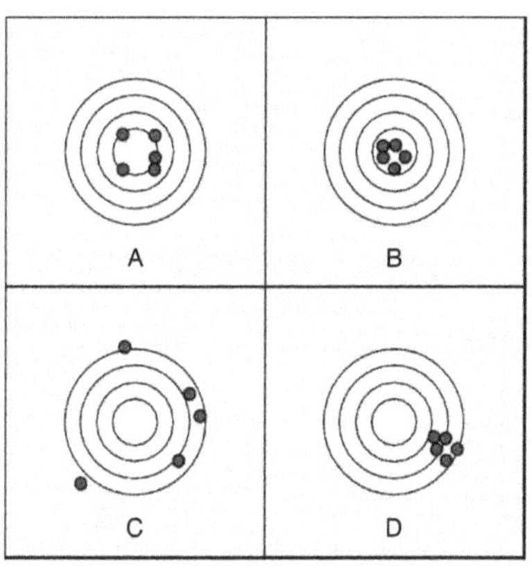

50. Which of the targets indicates good repeatability but poor bias?

   a. A

   b. B

   c. C

   d. D

51. Which of the targets indicates good repeatability, linearity, bias, and accuracy?

   a. A

   b. B

   c. C

   d. D

52. Which of the following is used to drill down into an issue to determine its underlying root cause?

   a. Affinity diagram

   b. MSA

   c. VSM

   d. 5 Whys

53. Check sheets are best used for which of the following?

   a. Systematically recording data for further analysis

   b. Identifying and narrowing the scope of a problem

   c. Developing and collecting a large number of ideas in a short time

   d. Organizing and displaying potential causes of an issue

Use the following table to answer questions 54–57.

	Current process				Action results			
Row	Severity	Occurrence	Detection	RPN	Severity	Occurrence	Detection	RPN
1	3	5	10	150	3	5	3	45
2	5	10	3	150	3	5	3	45
3	10	3	5	150	10	3	1	30

54. On which row of the FMEA should a process improvement team focus their efforts for the current state of the process?

    a. 1

    b. 2

    c. 3

    d. Since the RPNs are the same, no special effort is necessary.

55. Rank the order of risk for the current state of the process.

    a. 3, 2, 1

    b. 2, 3, 1

    c. 1, 2, 3

    d. Since the RPNs are the same, the risk is the same.

56. What is the overall percentage of RPN reduction in row 1?

    a. 30%

    b. 45%

    c. 55%

    d. 70%

57. What is the overall percentage of RPN reduction in row 3?

    a. 20%

    b. 30%

    c. 80%

    d. 85%

58. The primary purpose of the control plan is to:

    a. keep only one item as a variable and maintain all other characteristics as controls per experiment.

    b. implement validations on specific processes.

    c. ensure that process changes are maintained over time.

    d. create structure for how quality will be implemented throughout the process.

59. Sampling for a process step would most likely be found in which column of the control plan?
    a. Characteristic
    b. Specification
    c. Frequency
    d. Reaction plan

Use the following figure to answer questions 60–61.

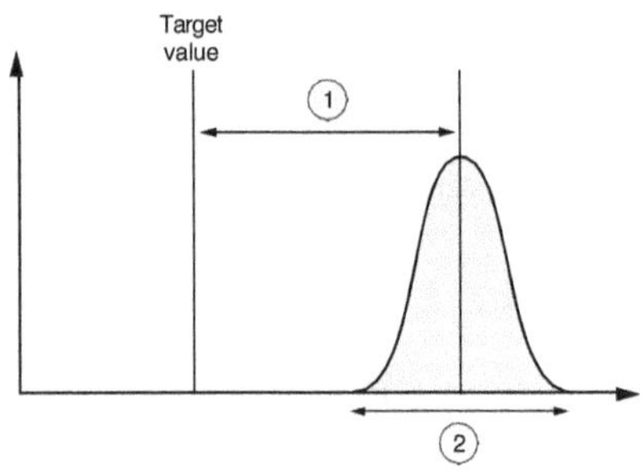

60. The encircled number 1 relates most closely to:
    a. the lower control limit.
    b. the upper control limit.
    c. accuracy.
    d. precision.
61. The encircled number 2 relates most closely to:
    a. the lower control limit.
    b. the upper control limit.
    c. accuracy.
    d. precision.
62. A tool used to monitor a process to make sure that the process improvements made are continuing to produce desired results over time is:
    a. the control plan.
    b. process audit.
    c. the control chart.
    d. GR&R.

63. At which stage of the 8D process are temporary, containment actions identified and implemented?

    a. D3

    b. D4

    c. D5

    d. D6

64. At which stage of the 8D process would a team modify a management system to prevent recurrence of a problem?

    a. D4

    b. D5

    c. D6

    d. D7

65. If a supplier's response to a SCAR identified the root cause as operator error, which of the following would be the most appropriate next step?

    a. Accept the supplier's response and close the SCAR.

    b. Visit the supplier to investigate and develop a better understanding of the root cause of the issue.

    c. Disqualify the supplier.

    d. Implement a tightened incoming inspection scheme.

66. Which of the following would be the most appropriate for use in qualifying a process with variable data?

    a. X-MR chart

    b. p chart

    c. c chart

    d. u chart

SOLUTIONS

1. c;

$$\text{Acceptance \%} = \frac{26(\text{ lots accepted })}{27(\text{ lots received })} = 96.30\%.$$

[IV.A.1]

2. a;

$$\text{Reject \%} = \frac{1(\text{lots rejected})}{27(\text{lots received})} = 3.70\%.$$

[IV.A.1]

3. c;

$$\frac{768\ (\text{total parts}) - 17\ (\text{rejected parts})}{768\ (\text{total parts})} = 97.79\%.$$

[IV.A.1]

4. c;

$$\text{OTD\%} = \frac{38\ (\text{orders delivered on time})}{40\ (\text{total orders delivered})} = 95.0\%.$$

[IV.A.1]

5. b;

$$\text{OTIF\%} = \frac{35\ (\text{orders delivered in full on time})}{40\ (\text{total orders delivered})} = 87.5\%.$$

[IV.A.1]

6. a; On time in full [OTIF] is used to measure the rate of orders that are fulfilled both on time and in full. [IV.A.1]
7. d; The risk-based Supplier Corrective Action Request (SCAR) decision matrix process follows the following sequence: identify the risk severity, determine the frequency, identify the severity-frequency rating, take appropriate action. [IV.A.1]
8. b; SCAR can be issued for audit findings and other supplier performance issues. [IV.A.1]
9. c; SCAR is the best formal tool used for communicating quality issues with a supplier. [IV.A.1]

10. c;

$$\text{Product quality} = \frac{25-3}{25} * 60 = 52.8.$$

[IV.A.2]

11. a;

$$\text{On-time delivery} = \frac{20}{25} * 40 = 32.0.$$

[IV.A.2]

12. c;

$$\text{Scorecard rating} = 52.8 + 32.0 = 84.8.$$

[IV.A.2]

13. d; Ship-to-stock programs use information regarding supplier performance and capabilities to determine whether materials from a supplier can bypass receiving inspection and be sent directly to stock. This may be done in an effort to implement just-in-time. A supplier performance index may be used as part of this evaluation, as could results from source inspections. [IV.A.2]

14. d; Lean manufacturing principles generally do not include the use of design of experiments (DOE) techniques. [IV.A.3]

15. b; Single-minute exchange of die (SMED) generally has a target time of 10 minutes. [IV.A.3]

16. d; When developing work instructions, the current best practice is to use photographs because visual instructions are often easier to understand and can help with training new people. [IV.A.3]

17. a; Many experts consider *safety* to be the sixth S in a 5S program. [IV.A.3]

18. c; *Kanban* is a Japanese word that means "visual signal" or "card." The signal is to perform a specific task or activity in a certain quantity. This technique is used in production and business processes that rely on a pull request or signal to do the work. [IV.A.3]

19. a; The true cost of inventory can be between 10% and 20% of the actual total dollar value at the component or piece part level, and up to 30% at the subassembly level. [IV.A.3]

20. c; The scope of work should be outlined or documented on a flowchart. [IV.A.3]

21. d; Procedures and work instructions require proper revision control as well as change management. [IV.A.3]

22. c; For general manufacturing, average inventory turns range from five to seven. Inventory on hand that is not being converted into sales is tying up cash that could be used for capital improvements or quality initiatives. [IV.A.3]

23. b; Kanban uses the production consumption rate to determine the right amount of inventory. Once that quantity is consumed, the signal is sent to the supplier to replenish the quantity required. [IV.A.3]

24. c; The 5S elements are:

    - *Seiri* (Sort). Eliminate whatever is not needed.
    - *Seiton* (Straighten). Organize whatever remains.
    - *Seiso* (Shine). Clean the work area.
    - *Seiketsu* (Standardize). Schedule regular cleaning and maintenance.
    - *Shitsuke* (Sustain). Make 5S a way of life. [IV.A.3]

25. b; *SMED* is a system used to reduce changeover time and improve timely response to demand. SMED is a series of techniques pioneered by Shigeo Shingo to facilitate changeovers of production machinery in less than 10 minutes. [IV.A.3]

26. c; Taiichi Ohno, an industrial engineer at Toyota, developed kanban to improve manufacturing efficiency. [IV.A.3]

27. b; *Kaizen* is a Japanese term that means gradual unending improvement by doing little things better and setting and achieving increasingly higher standards. [IV.A.3]

28. b; The takt time to produce 1000 quarters due in five days with one eight-hour shift per day is calculated by:

$$\text{Takt time} = \frac{\text{Time available}}{\text{Number of items to be produced}} = \frac{8\frac{\text{hrs}}{\text{day}} * 60\frac{\text{minutes}}{\text{hour}} * 5\text{ days}}{1000\text{ units}}$$
$$= 2.4 \text{ minutes per unit.}$$

    [IV.A.3]

29. b; Poka-yoke comes from Japan, and is also known as *error-proofing*. It is a method of preventive action—a technique used to prevent errors from occurring. [IV.A.3]

30. a; Overproduction is producing more of a product than customer demand requires. In this case, the next process is the internal customer. [IV.A.3]

31. c; Standardized work means that every step or activity should be done the same way. Standardized work helps to reduce process variation and provides a more consistent product or service. [IV.A.3]

32. c; The goal of a lean system is to eliminate waste from a process. The first step is to identify opportunities for improvement, which can be found using a value stream map. Value stream maps are a visual representation of all the steps in a process, including information about timing. Value stream maps also identify the value-added and non-value-added activities in a process. [IV.A.3]

33. a; The 5S elements are:

    - *Seiri* (Sort). Eliminate whatever is not needed
    - *Seiton* (Straighten). Organize whatever remains
    - *Seiso* (Shine). Clean the work area
    - *Seiketsu* (Standardize). Schedule regular cleaning and maintenance
    - *Shitsuke* (Sustain). Make 5S a way of life. [IV.A.3]

34. d; The purpose of a material review board (MRB) is to determine the appropriate disposition for nonconforming components that are discovered. The MRB may also determine any additional actions to take to prevent similar nonconformities in future. [IV.B]

35. b; Material segregation ensures that nonconforming products are not mixed with conforming products. Segregation can be physical, electronic using an Enterprise Resource Planning (ERP) system, or a combination of both. [IV.B]

36. b; The poor battery life of the cell phone reduces the usability of the product but will not likely lead to an injury or significant economic loss. Therefore, the company would classify this as a major defect. [IV.B]

37. d; The 15 randomly selected PCBs failed the "burn-in" test. These 15 PCBs were immediately labeled, placed on electronic hold, and moved to the quarantine storage area. The MRB should first ensure that the remaining 85 PCBs are placed in quarantine per procedure. [IV.B]

38. c; Since these PCBs are critical to the product being manufactured, the MRB should place all PCBs from the previous lot on hold. [IV.B]

39. d; The best course of action for the MRB would be to perform a retest of all of the equipment that was assembled using the PCBs from the skip-lot and release. [IV.B]

40. c; Since there is an assignable cause associated with the lot of PCBs that failed incoming inspection, the most reasonable action for the MRB would be to retest based on an Acceptance Quality Limit (AQL) sampling of the equipment that was assembled using the PCBs from the skip-lot and release. [IV.B]

41. a; Executing a *supplier change notification* agreement is the best course of action. This would allow the company to evaluate whether the change could have a potential adverse effect. [IV.B]

42. b; A cause-and-effect diagram (also called the *Ishikawa diagram* or *fishbone diagram*) traditionally divides causes into several generic categories. In use, a large empty diagram is often drawn on a whiteboard or flip chart to collect and visually display potential causes of a problem. [IV.C.1]

43. a; Pareto charts can be used to help a process improvement team focus their efforts based on the Process Failure Mode and Effects Analysis (PFMEA). [IV.C.1]

44. a; Fault trees provide a visualization of hierarchical relationships of events that lead to failure (or other undesirable outcome). The team here wishes to determine potential actions that can lead to process failures. [IV.C.1]

45. a; Figures A and B do not indicate the presence of an interaction because the lines are parallel.

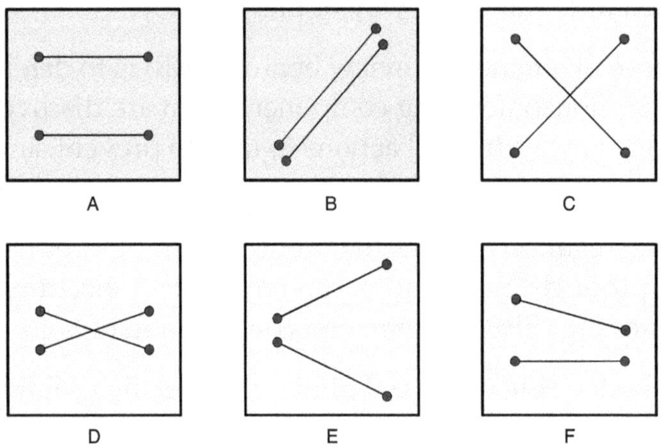

[IV.C.1]

46. d; Figures C and D indicate the most significant interactions because the lines intersect. Figures E and F also indicate the presence of interaction. However, the interaction is not as significant as in Figures C and D.

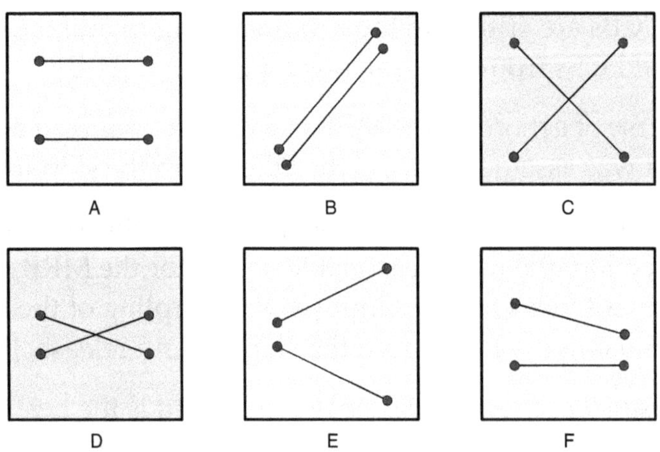

[IV.C.1]

47. a; Brainstorming is a group process used to generate ideas in a nonjudgmental way. The purpose of brainstorming is to generate a large number of ideas about an issue. [IV.C.1]

48. c; The affinity diagram is a tool used to organize information and help achieve order out of the chaos that can develop in a brainstorming session. Large amounts of data, concepts, and ideas are grouped based on their natural relationships to one another. [IV.C.1]

49. b; *Linearity* is the difference in bias through the operating range of measurements. A measurement system that has good linearity will have a constant bias no matter the magnitude of measurement. [IV.C.1]

50. d; This grouping shows good repeatability (precision) and linearity, but low accuracy due to poor bias.

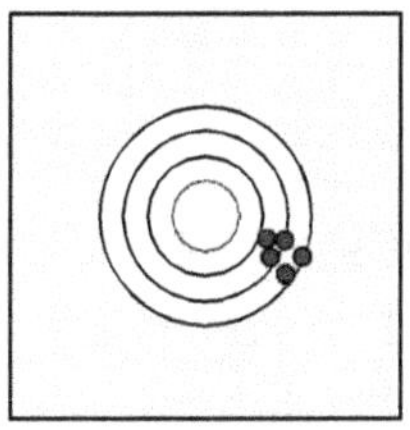

[IV.C.1]

51. b; This grouping shows good repeatability (precision), linearity, bias, and accuracy.

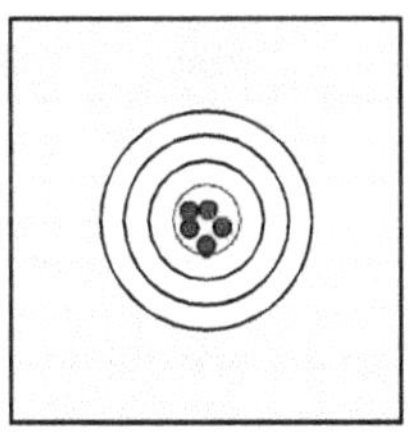

[IV.C.1]

52. d; 5 Whys are used to drill down into an issue to uncover the underlying root cause. The other options are tools used elsewhere in the DMAIC process. [IV.C.1]

53. a; Check sheets are data collection devices. The other options describe Is/Is Not analysis, brainstorming, or cause-and-effect diagrams. [IV.C.1]

54. c;

	Current process				Action results			
Row	Severity	Occurrence	Detection	RPN	Severity	Occurrence	Detection	RPN
1	3	5	10	150	3	5	3	45
2	5	10	3	150	3	5	3	45
3	10	3	5	150	10	3	1	30

The process improvement team should first focus their efforts on the current state of the process for row 3, as row 3 has the highest severity (10). [IV.C.2]

55. a;

	Current process				Action results			
Row	Severity	Occurrence	Detection	RPN	Severity	Occurrence	Detection	RPN
1	3	5	10	150	3	5	3	45
2	5	10	3	150	3	5	3	45
3	10	3	5	150	10	3	1	30

The order of risk for the current state of the process is row 3, row 2, row 1. All three rows have the same Risk Priority Number (RPN). Based on the information provided, the risk priority will be determined by the severity, then by the occurrence. [IV.C.2]

56. d;

	Current process				Action results			
Row	Severity	Occurrence	Detection	RPN	Severity	Occurrence	Detection	RPN
1	3	5	10	150	3	5	3	45
2	5	10	3	150	3	5	3	45
3	10	3	5	150	10	3	1	30

The percentage of RPN reduction in row 1 is calculated by the formula:

$$\%\ \text{RPN reduction} = \frac{\text{RPN}_i - \text{RPN}_r}{\text{RPN}_i} = \frac{150-45}{150} = 0.70 \text{ or } 70\%$$

where RPN_i = Initial RPN and RPN_r = Revised RPN.
[IV.C.2]

57. c;

	Current process				Action results			
Row	Severity	Occurrence	Detection	RPN	Severity	Occurrence	Detection	RPN
1	3	5	10	150	3	5	3	45
2	5	10	3	150	3	5	3	45
3	10	3	5	150	10	3	1	30

The percentage of RPN reduction in row 3 is calculated by the formula

$$\%\ \text{RPN reduction} = \frac{\text{RPN}_i - \text{RPN}_r}{\text{RPN}_i} = \frac{150-30}{150} = 0.80 \text{ or } 80\%$$

where RPN_i = Initial RPN and RPN_r = Revised RPN.
[IV.C.2]

58. d; A *control plan* is a living document that identifies critical input and output variables and associated activities that must be performed to maintain control of the variation of processes, products, and services in order to minimize deviation from their preferred values. A control plan is defined as a living document; it is designed to be maintained. [IV.C.2]

59. c; The control plan should specify a sampling plan that experience has shown to be effective in detecting changes to the process. The plan should identify gages or fixtures to be used. If measurements are to be made, the plan should cross-reference a control plan for monitoring GR&R for the gage involved. [IV.C.2]

60. c; *Accuracy* refers to how closely the values are centered relative to the target/nominal value. [IV.C.2]

61. d; *Precision* refers to how closely the values are grouped or clustered together. [IV.C.2]

62. c; Once a process has been improved, it must be monitored to ensure that the gains are maintained and to determine when additional improvements are required. Control charts are used to monitor the stability of the process and determine when a special cause is present and when to take appropriate action. While a process audit can also be used as a tool to evaluate a process, it provides more of a snapshot view, while control charts serve to monitor a process over time. [IV.C.2]

63. a; Temporary containment actions are chosen in stage D3: Develop interim containment plan and implement and verify interim actions. [IV.C.2]

64. d; Modification of a management system to prevent recurrence of a problem is done in stage D7: Take preventive measures. [IV.C.2]

65. b; Human error is rarely a true root cause—dig deeper to find the underlying process issue that allowed the root cause to occur. Tightened inspection should have already been implemented if performance has degraded enough that a SCAR was warranted. The supplier may eventually be disqualified if a pattern of poor performance continues. [IV.C.2]

66. a; X-moving range (X-MR) is the only choice of those presented that is appropriate for variable data. The others are charts for attribute data. [IV.C.2]

Section V

Supplier Quality Management

(68 questions)

A. SUPPLIER QUALITY MONITORING

1. *Supplier audit.* Apply the stages of a quality audit, including audit planning, conducting the initial audit, and executing periodic reevaluation. Understand and apply the various types of quality audits (e.g., product, process, and management system) and audit methods (e.g., virtual, on-site, and desktop). (Apply)
2. *Audit reporting and follow-up.* Apply and analyze process audit reporting and follow up, including verification of the effectiveness of corrective action. (Analyze)
3. *Supplier communication.* Evaluate various communication techniques such as periodic reviews, metric and performance indices, change management, notifications, recalls, change requests, and business updates. Maintain active communication with suppliers to assess risk and take appropriate action. (Evaluate)
4. *Supplier development and remediation.* Identify and analyze present and future training needs and gaps, using quality methods and tools, such as kaizen and benchmarking. Use process improvement tools such as DMAIC, cycle time reduction, defect rate, and cost reduction. Evaluate supplier remediation to develop and manage improvement plans. (Evaluate)
5. *Project management basics.* Understand and apply various types of project reviews, such as phase-end, management, and retrospectives or post-project reviews to assess project performance and status, review issues and risks, and discover and capture lessons learned from the project. Apply forecasts, resources, schedules, and task and cost estimates to develop and monitor project plans. (Apply)

B. TEAMS AND TEAM PROCESSES

1. *Team development.* Identify and describe the various types of teams and the classic stages of team development: forming, storming, norming, performing, and adjourning. (Apply)
2. *Team roles.* Define and describe various team roles and responsibilities for leader, facilitator, coach, and individual member. (Understand)
3. *Performance and evaluation.* Describe various techniques to evaluate training, including evaluation planning, feedback surveys, pre-training testing, and post-training testing. (Understand)

C. COMPLIANCE WITH REQUIREMENT AND SUPPLIER CATEGORIZATION

Understand and evaluate compliance with regulations and industry standards (e.g., RoHS, Governmental regulatory authorities, and ISO), specifications, contracts, agreements, and certification authority. Evaluate and categorize suppliers based on risk and performance. (Evaluate)

QUESTIONS

1. All of the following are customers of an audit *except:*
   a. the person responsible for the activity being audited.
   b. upper management.
   c. other functions affected by the activity being audited.
   d. the lead auditor.
2. Which of the following is an example of a first-party audit?
   a. An ISO registration audit
   b. A departmental audit
   c. A dock audit
   d. A supplier audit

3. Which of the following is an example of a second-party audit?
   a. An ISO registration audit
   b. A departmental audit
   c. A dock audit
   d. A supplier audit
4. Which of the following is an example of a third-party audit?
   a. An ISO registration audit
   b. A departmental audit
   c. A dock audit
   d. A supplier audit

Use the following information to answer questions 5–7.

An audit is being performed, specifically focusing on documentation (SOPs and WIs) and records. The documentation policy requires that the "same function" that originally approved the document must approve changes to documents. The procedure also requires that the documents be available to the employees electronically. The records policy requires the record to be legible, protected from unintended alterations, and with electronic scans acceptable as the permanent record. During the audit, the auditor observed the following:

> *SOP 110 Preventive Maintenance—The procedure was originally approved on 10/1/2017 by Jane Doe, Maintenance Manager, and Bill Smith, Quality Manager.*
>
> *SOP 115 Production Control—The procedure was originally approved on 11/12/2016 by Todd Jones, Production Manager, and Bill Smith, Quality Manager.*
>
> *Production Batch Record 1234 was scanned and saved on the company server, which is cloud-based. The server requires a unique username and password for access. The batch was approved for release on 3/14/2018.*

5. SOP 110 Preventive Maintenance needs to be updated; however, Jane Doe, Maintenance Manager, was promoted to director of operations, and a replacement was recently hired. Which of the following individual(s) should approve the change to the procedure?

   a. Only the quality manager

   b. The quality manager and the maintenance manager

   c. The director of operations and the maintenance manager

   d. The director of operations, the maintenance manager, and the quality manager

6. SOP 115 Production Control needs to be updated; however, Todd Jones, Production Manager, is on vacation for six weeks. The production manager directly reports to the director of operations. Which of the following individual(s) should approve the change to the procedure?

   a. The director of operations and the quality manager

   b. The director of operations and the maintenance manager

   c. Only the quality manager

   d. The procedure should not be changed until production manager Todd Jones returns from vacation

7. The auditor noticed that Production Batch Record 1234 was approved for release on 3/14/2018, scanned, and saved to the cloud-based server. The auditor asked for the original paper copy. The director of operations informed the auditor that under the procedure, scanned electronic copies are acceptable as the permanent record. Is this a violation of the procedure?

   a. Yes, original paper documents should never be destroyed.

   b. Yes, the cloud-based server is not physically located on the company property.

   c. No, paper copies are required to be saved.

   d. No, the procedure allows electronic scans as the permanent record.

8. What type of audit evaluates final "fitness for use"?

   a. Product audit

   b. System audit

   c. Process audit

   d. Desk audit

9. What type of audit is best suited for verifying results and effectiveness of continuous improvement actions?
   a. Product audit
   b. System audit
   c. Process audit
   d. Desk audit

10. What type of audit is best suited to verify the effectiveness of the QA program?
   a. Product audit
   b. System audit
   c. Process audit
   d. Desk audit

11. A virtual audit may be appropriate for which of the following?
   a. Interviewing people and reviewing documents
   b. Reviewing a supplier's manufacturing process
   c. Source inspection of a capital equipment purchase at the supplier prior to shipment
   d. A full management system audit

12. A company sends a supplier a questionnaire to complete and return regarding the basic structure of the supplier's quality management system. This is an example of which of the following types of audit?
   a. Virtual audit
   b. Process audit
   c. Desktop audit
   d. Product audit

13. The following are appropriate reasons for a quality audit *except*:
   a. disciplinary control.
   b. product quality evaluation.
   c. preventive action.
   d. qualifying a new supplier.

14. Which of the following provides guidance on auditing management systems?
    a. ISO 9001
    b. ISO 13485
    c. ISO 15378
    d. ISO 19011
15. Which of the following is responsible for the audit report's content, accuracy, and timely publication?
    a. Audit team
    b. Sourcing/purchasing lead
    c. Lead auditor
    d. Supplier quality manager
16. If the time gap between conducting an audit and issuing the audit report is excessive, the auditee may get the unintended message that:
    a. auditors are overwhelmed.
    b. the audit and the corrective action on its findings are not a priority.
    c. the audit was a paper exercise.
    d. they did well during the audit.
17. A supplier corrective action can be considered closed when:
    a. the plan is submitted to the customer.
    b. purchasing and supplier quality approve.
    c. it has been implemented and verified.
    d. it has been filed with the auditor.
18. The best person to review and approve a supplier audit response is:
    a. the supplier quality manager.
    b. the director of procurement.
    c. the supplier's management team.
    d. the auditor.

19. Strategies for optimizing overall quality cost is a topic for:
    a. joint economic planning.
    b. a quarterly business review.
    c. a supplier scorecard.
    d. management review.
20. Total dollar value purchased (spend), % defective dollars versus spend, % of lots rejected, and composite supplier rating scores are all examples of:
    a. a quarterly business review (QBR) agenda.
    b. financial due diligence.
    c. supplier feedback topics.
    d. supplier selection criteria.

Questions 21–24 concern different types of supplier assessment metrics. Choose the category based on the list offered.

21. Percent over-orders, percent under-ordered, and percent early delivery are examples of:
    a. timeline metrics.
    b. quality metrics.
    c. delivery (JIT) metrics.
    d. compliance metrics.
22. Percent orders missing certificates and percent of reported required quality information are examples of:
    a. timeline metrics.
    b. quality metrics.
    c. delivery (JIT) metrics.
    d. compliance metrics.
23. Percent defective, ppm, DPMO, and percent nonconforming are examples of:
    a. cost metrics.
    b. quality metrics.
    c. delivery (JIT) metrics.
    d. compliance metrics.

24. Dollars rejected versus purchased is an example of a:

    a. cost metric.

    b. quality metric.

    c. delivery (JIT) metric.

    d. compliance metric.

25. Document control and configuration control are key aspects of which of the following?

    a. Change management

    b. CAPA

    c. Product audit

    d. Inspection plans

26. A company's management researches the company's competitive position through comparisons with other companies' products and services. This is an example of which of the following?

    a. Kaizen

    b. Benchmarking

    c. CAPA

    d. DMAIC

27. Incremental continuous improvement is also known as:

    a. kaizen.

    b. Six Sigma.

    c. lean enterprise.

    d. PDCA.

28. Excess motion, repairs/rework, waiting, and excess inventory are some examples of:

    a. steps in a process.

    b. muda.

    c. cycle time reduction.

    d. steps to take to ensure material is ready when production is.

29. Efforts to improve processes to reduce waste, increase throughput, and increase capacity are all indicative of:

    a. cycle time reduction.

    b. root cause analysis.

    c. DMAIC.

    d. control plans.

30. Which of the following represents costs associated with measuring, evaluating, or auditing products or services to ensure conformance to quality standards and performance requirements?

    a. Prevention costs

    b. Appraisal costs

    c. Internal failure costs

    d. External failure costs

31. Which of the following is generally the costliest category of quality costs?

    a. Prevention costs

    b. Appraisal costs

    c. Internal failure costs

    d. External failure costs

32. Supplier remediation activities include all of the following *except*:

    a. PPAP instruction.

    b. joint kaizen events.

    c. technical assessments.

    d. price negotiations.

33. SMART goals—This method helps ensure that the goals have been fully investigated and provides a way to clearly understand the implications of the goal-setting process. What does the M refer to?

    a. Meaningful

    b. Management

    c. Metrics

    d. Measurable

34. CLEAR goals are a newer method for setting goals that takes into consideration the environment of today's fast-paced business. What does the L refer to?

    a. Lesser

    b. Larger

    c. Limited

    d. Leveraged

35. What type of chart provides a visual timeline that can be used to plan tasks and visualize the project timeline?

    a. Pareto chart

    b. Gantt chart

    c. Flowchart

    d. Control chart

36. What type of plan is used to identify unrealistic time and cost estimates, customer review cycle, budget cuts, changing requirements, and lack of committed resources?

    a. Risk management

    b. Communication

    c. Budget

    d. Control

37. ______ is a visual representation that breaks down the project scope into manageable sections for the team.

    a. DMAIC

    b. PDCA

    c. WBS

    d. VSM

38. _____ identify high-level goals that need to be met throughout the project. These are included in the Gantt chart.

    a. Metrics

    b. Measurables

    c. Specifications

    d. Milestones

39. Goals are normally set in the planning stage. Which of the following tools is best used to describe project goals?

    a. PDCA

    b. CLEAR

    c. SMART

    d. (b) and (c)

40. The team phase where team members struggle to understand the team goal and its meaning for them individually is referred to as:

    a. norming.

    b. forming.

    c. performing.

    d. storming.

41. The team phase where team members begin to understand the need to operate as a team rather than as a group of individuals is referred to as:

    a. storming.

    b. forming.

    c. norming.

    d. performing.

42. The team phase where team members understand one another and recognize one another's strengths and weaknesses is referred to as:

    a. storming.

    b. performing.

    c. norming.

    d. forming.

43. What are the five stages of team development, in order?

    a. Forming, norming, storming, performing, adjourning

    b. Storming, adjourning, norming, forming, performing

    c. Forming, storming, norming, performing, adjourning

    d. Storming, performing, forming, norming, adjourning

44. How can a team best overcome anxiety during the forming stage?

    a. Create a team name and slogan to promote unity.

    b. Let members pick their own teams to promote harmony.

    c. Determine a team leader quickly to ensure timely actions.

    d. Set clear goals, timelines, and roles to communicate expectations.

45. This audit team role is responsible for calling the meetings, making meeting arrangements, running the meetings, and reporting progress to the client and the auditee.

    a. Sponsor

    b. Team members

    c. Facilitator

    d. Lead auditor

46. This audit team role ensures that everyone understands his or her assignment before the audit starts.

    a. Client

    b. Auditor

    c. Facilitator

    d. Lead auditor

47. This audit team role should provide the necessary resources for the team members, such as forms, examples, and working papers, and if necessary, instruct team members on the use of working papers.

    a. Client

    b. Auditor

    c. Facilitator

    d. Lead auditor

48. Audit teams are ________ teams formed to carry out the audit plan and achieve the supplier audit objectives.

    a. ad hoc

    b. administrative

    c. specialized

    d. complicated

49. The need for facilitators and facilitation techniques ________ with team member training, experience, and successful team history.

    a. increases

    b. decreases

    c. remains the same

    d. is not considered

50. A key role of a project ________ is to track team activities and remove obstacles to team progress.

    a. leader

    b. consultant

    c. facilitator

    d. sponsor

51. The team facilitator asks each team member to suggest one idea during each round of a brainstorming session. This method can be used when:

    a. the meeting session is over two hours.

    b. the team is behind schedule.

    c. some team members are dominating the discussion.

    d. management is observing the team.

52. Which of the following is a characteristic of a team that is effectively functioning?

    a. A sense of team interdependence

    b. A clear sense of purpose

    c. A focus on individual tasks

    d. A knowledge of company policies

53. A team facilitator can promote positive team relationships by:

    a. recording meeting notes.

    b. debating each item.

    c. keeping track of time.

    d. encouraging participation.

54. What can occur when team members put more value on getting along than on achieving team goals?
    a. Groupthink
    b. Conflict
    c. Synergy
    d. Collaboration

55. Which characteristic best indicates a team member's ability to contribute to a project team?
    a. Leadership skills
    b. Ability to conform
    c. Knowledge of the problem
    d. Social justification

56. Which of the following European Union (EU) regulations restricts the use of certain chemicals, including lead (Pb), mercury (Hg), and cadmium (Cd), in electrical and electronic products?
    a. REACH
    b. RoHS
    c. OSHA
    d. Conflict minerals

57. This is the most common type of specifications. It defines what is required for a product to perform as expected by the consumer.
    a. Analytical specifications
    b. Product specifications
    c. Process specifications
    d. Raw material specifications

58. This type of specifications defines the parameters of the manufacturing process that must be controlled to produce a product.
    a. Quality management specifications
    b. Product specifications
    c. Process specifications
    d. Raw material specifications

59. Where applicable, this type of specifications defines the testing and measurement methodologies for measuring a required level of accuracy.
    a. Quality management specifications
    b. Product specifications
    c. Analytical specifications
    d. Raw material specifications
60. Which of these basic types of specifications is generally determined first?
    a. Product specifications
    b. Process specifications
    c. Raw material specifications
    d. Analytical specifications
61. This type of specifications defines the management practices under which you wish to have products produced.
    a. Product specifications
    b. Quality management specifications
    c. Analytical specifications
    d. Raw material specifications
62. Which of the following controls the export and import of defense-related articles and services on the United States Munitions List (USML)?
    a. DMAIC
    b. REACH
    c. ITAR
    d. Conflict minerals
63. Conflict minerals are natural minerals extracted in a conflict zone and sold to perpetuate the fighting. Conflict minerals are associated with which geographic region?
    a. Middle East
    b. Antarctica
    c. North America
    d. Democratic Republic of the Congo

64. Which of these is among the most commonly mined conflict minerals?
    a. Tin
    b. Aluminum
    c. Copper
    d. Lead

65. Agreements regarding change control requirements and mechanisms are generally included in which of the following?
    a. Quality agreements
    b. Quality management specifications
    c. Process specifications
    d. Audit plans

66. Which of these are the three main components of risk assessment?
    a. Identification, management, and control
    b. Categorization, analysis, and mitigation
    c. Identification, analysis, and evaluation
    d. Categorization, management, and monitoring

67. A company rates its raw material suppliers based on two equally important key risk factors, nonconformance rate and incoming inspection reject rate. The ratings for the past quarter are included in the following table. Based on this information, which of these suppliers represents the largest risk to the company?

Supplier	Nonconformance Rate	Incoming Inspection rate
A	3	1
B	3	5
C	1	5
D	3	3

    a. A
    b. B
    c. C
    d. D

68. Based on the following risk matrix table stratifying a company's suppliers, which supplier represents the highest level of risk to the company?

	Parts Criticality—Low	**Parts Criticality—High**
Spend Level—Low	A	B
Spend Level—High	C	D

a. A

b. B

c. C

d. D

SOLUTIONS

1. d; The options presented in (a), (b), and (c) are ultimately customers of an audit. [V.A.1]

2. b; A first-party audit is performed within an organization to measure its strengths and weaknesses against its own procedures or methods and/or against external standards adopted by (voluntary) or imposed on (mandatory) the organization. A first-party audit is an internal audit conducted by auditors who are employed by the organization being audited but who have no vested interest in the audit results of the area being audited. [V.A.1]

3. d; A second-party audit is an external audit performed on a supplier by a customer or by a contracted organization on behalf of a customer. A contract is in place, and the goods or services are being, or will be, delivered. Second-party audits are subject to the rules of contract law, as they provide contractual direction from the customer to the supplier. Second-party audits tend to be more formal than first-party audits because audit results could influence the customer's purchasing decisions. [V.A.1]

4. a; A third-party audit is performed by an audit organization independent of the customer–supplier relationship and free of any conflict of interest. Independence of the audit organization is a key component of a third-party audit. Third-party audits may result in certification, registration, recognition, an award, license approval, a citation, a fine, or a penalty issued by the third-party organization or an interested party. [V.A.1]

5. b; SOP 110 Preventive Maintenance needs to be updated; however, Jane Doe, Maintenance Manager, was promoted to director of operations, and a replacement was recently hired. The procedure was originally approved on 10/1/2017 by maintenance manager Jane Doe and Bill Smith, Quality Manager. The procedure should be signed by the same function, not necessarily the same individual. Therefore, the maintenance manager and the quality manager should both approve the changes to the procedure. [V.A.1]

6. a; SOP 115 Production Control Documentation needs to be updated; however, Todd Jones, Production Manager, is on vacation for six weeks. Since the production manager directly reports to the director of operations, it would be appropriate for the director of operations and the quality manager to approve the changes to the document. [V.A.1]

7. d; The auditor noticed that Production Batch Record 1234 was approved for release on 3/14/2018, scanned, and saved to the cloud-based server. The auditor asked for the original paper copy. The director of operations informed the auditor that under the procedure, scanned electronic copies are acceptable as the permanent record. This would not constitute a violation of the procedure. However, if the scan copy were not legible, it could be a violation of the procedure. [V.A.1]

8. a; A *product audit* is an examination of a particular product or service (hardware, processed material, software) to evaluate whether it conforms to requirements (specifications, performance standards, and customer requirements). [V.A.1]

9. c; A *process audit* is a verification that processes are working within established limits. It evaluates an operation or method against predetermined instructions or standards to measure conformance to these standards and the effectiveness of the instructions. [V.A.1]

10. b; A *system audit* is an audit conducted on a management system. It can be described as a documented activity performed to verify, by examination and evaluation of objective evidence, that applicable elements of the system are appropriate and effective and have been developed, documented, and implemented in accordance and in conjunction with specified requirements. [V.A.1]

11. a; Virtual audits are appropriate for conducting interviews and for reviewing documents, assuming that the documents can be shared electronically. The other scenarios would most appropriately be conducted using on-site audits. [V.A.1]

12. c; A *desktop audit* is when the auditor or sponsoring company sends a questionnaire to the auditee to complete and return. [V.A.1]

13. a; Disciplinary control should never be the reason for a quality audit. Quality audits are used to assess systems, processes, and products, and seek improvement opportunities. [V.A.1]

14. d; ISO 19011 provides guidance on auditing management systems, including the principles of auditing, managing an audit program, and conducting management system audits, as well as guidance on the evaluation of the competence of individuals involved in the audit process, including the person managing the audit program, auditors, and audit teams. [V.A.2]

15. c; The lead auditor is ultimately responsible for the audit reporting, including accuracy, completeness, and timely publication. [V.A.2]

16. b; Any findings important enough to document during an audit are also urgent to resolve. Untimely reporting sends the message that there are more important and urgent priorities. [V.A.2]

17. c; Closure can only occur after implementation and verification. Approval of plans or just customer approval are not enough. Objective evidence of implementation is required. A follow-up audit might be a good idea as well to verify effectiveness. [V.A.2]

18. d; Only the auditor has the in-depth first-person knowledge to appropriately judge the acceptability of an audit finding response from the supplier. [V.A.2]

19. a; Quality costs strategy is a staple of joint economic planning, along with discussing the value proposition. Actual cost of poor quality is a great category for scorecards and a topic for a Quarterly Business Review (QBR). [V.A.3]

20. c; These are all excellent topics to include on supplier feedback reports to enhance a strong customer–supplier relationship. They could be QBR agenda items as well, but QBR is just one type of supplier feedback. [V.A.3]

21. c; Although timeliness and delivery are similar, these are delivery metrics. They deal with just-in-time delivery. [V.A.3]

22. d; Although quality and compliance are similar, these are compliance metrics because they refer to required documentation. [V.A.3]

23. b; Although quality and compliance are similar, these are quality metrics because they refer to product quality and meeting/not meeting requirements. [V.A.3]

24. a; Since dollar value is mentioned, cost is the right choice. [V.A.3]

25. a; Document control and configuration management are key aspects of change management. Product audit and inspection plans may reference aspects of document control and configuration management but do not describe how these are managed. Corrective and Preventive Action (CAPA) may be necessary if document control and configuration management are not managed properly. [V.A.3]

26. b; Performance benchmarking enables management to assess a company's competitive position through product and service comparisons. This activity may be used as part of the approach for the other choices listed. [V.A.4]

27. a; Kaizen refers to incremental improvement that never stops. The other choices are great improvement tools, but are not necessarily used for iterative, ongoing improvement. [V.A.4]

28. b; These are verbatim examples of waste, or muda. [V.A.4]

29. a; Cycle time is the most logical answer. The other choices are tools or approaches that may be used to achieve cycle time reduction. [V.A.4]

30. b; This is the definition of appraisal costs. The others are also quality costs but are of different types. [V.A.4]

31. d; Since external failure costs encompass the costs associated with escapes, they are the costliest category of quality costs. [V.A.4]

32. d; (a), (b), and (c) are great areas in which to resolve performance issues with the supplier. [V.A.4]

33. d;

S	Specific
M	Measurable
A	Attainable
R	Realistic
T	Timely

[V.A.5]

34. c;

C	Collaborative
L	Limited
E	Emotional
A	Appreciable
R	Refinable

[V.A.5]

35. b; Gantt charts provide a graphical timeline that can be used to plan tasks and visualize the project timeline. [V.A.5]

36. a; A risk management plan can be used to identify *all* potential risks. These include unrealistic time and cost estimates, customer review cycle, budget cuts, changing requirements, and lack of committed resources. [V.A.5]

37. c; Work breakdown structure (WBS) is a visual representation that breaks down the project scope into manageable sections for the team. [V.A.5]

38. d; The milestones identify high-level goals that need to be met throughout the project. These are included in the Gantt chart. [V.A.5]

39. d; SMART and CLEAR tools can be used to describe project goals. [V.A.5]

40. b; Team *forming* is when members struggle to understand the team goal and its meaning for them individually. [V.B.1]

41. c; Team *norming* is when members begin to understand the need to operate as a team rather than as a group of individuals. [V.B.1]

42. b; Team *performing* is when team members work together to reach their common goal. [V.B.1]

43. c; The five stages of team development, in order, are forming, storming, norming, performing, and adjourning. [V.B.1]

44. d; Setting clear goals, timelines, and roles can help a team overcome anxiety during the forming stage. [V.B.1]

45. d; The lead auditor is responsible for calling the meetings, making meeting arrangements, running the meetings, and reporting progress to the client and the auditee. The lead auditor should take ownership of the audit process. The lead auditor normally controls the team output or deliverables. Lead auditors must be able to guide the team and make decisions as needed to ensure that the team is effective. The lead auditor should be a qualified (competent) auditor. [V.B.2]

46. d; The lead auditor ensures that everyone understands his or her assignment before the audit starts. [V.B.2]

47. d; The lead auditor should provide the necessary resources for the team members, such as forms, examples, and working papers, and if necessary, instruct team members on the use of the working papers. [V.B.2]

48. a; *Audit teams* are ad hoc teams formed to carry out the audit plan and achieve the supplier audit objectives. [V.B.2]

49. b; The need for facilitators and facilitation techniques decreases with team member training, experience, and successful team history. [V.B.2]

50. d; The project sponsor's chief roles are to provide the scope of the project, provide the project resources, review and track team activities, and remove obstacles to team progress. [V.B.3]

51. c; When a team member(s) is/are dominating a discussion, the team facilitator can ask each team member to suggest one idea during individual rounds of a brainstorming session. This will help solicit participation from the larger group. [V.B.3]

52. b; A characteristic of a team that is effectively functioning is a clear sense of purpose. [V.B.3]

53. d; By encouraging participation, the team facilitator can promote positive team relationships. [V.B.3]

54. a; Groupthink can occur when team members put more value on getting along than on achieving team goals. [V.B.3]

55. c; Knowledge of the problem is an indicator of a team member's ability to contribute to a project team. [V.B.3]

56. b; The Restriction of Hazardous Substances (RoHS) Directive restricts the types of hazardous materials that can be used in electrical and electronic products. Under RoHS, the following substances are prohibited from entering the market for electrical and electronic products: lead (Pb), mercury (Hg), cadmium (Cd), hexavalent chromium (Cr VI), polybrominated biphenyls (PBB), polybrominated diphenyl ethers (PBDE), and four different phthalates (DEHP, BBP, DBP, and DIBP). [V.C]

57. b; Product specifications are the most common type of specifications. They define what is required for a product to perform as expected by the consumer. [V.C]

58. c; Process specifications define the parameters of the manufacturing process that must be controlled in order to produce a product. [V.C]

59. c; Analytical specifications define the analytical (testing and measurement) methodologies for measuring a required level of accuracy. [V.C]

60. a; Product specifications are generally determined first, as they are the drivers for the other types of specifications. [V.C]

61. b; Quality management specifications define the management practices under which you wish to have products produced. [V.C]

62. c; International Traffic in Arms Regulations (ITAR) controls the export and import of defense-related articles and services on the United States Munitions List (USML). ITAR does not apply to general scientific, mathematical, or engineering principles that are taught in schools or are in the public domain. [V.C]

63. d; Conflict minerals are natural minerals extracted in a conflict zone and sold to perpetuate the fighting. Conflict minerals are associated with the Democratic Republic of the Congo. The four most commonly mined conflict minerals are cassiterite (for tin), wolframite (for tungsten), coltan (for tantalum), and gold ore. [V.C]

64. a; The four most commonly mined conflict minerals are cassiterite (for tin), wolframite (for tungsten), coltan (for tantalum), and gold ore. [V.C]

65. a; Quality agreements usually contain language around change managements, especially written change notifications and evaluation of any proposed change. [V.C]

66. c; The three main components of risk assessment are identification, analysis, and evaluation. Management, mitigation, control, and monitoring are aspects of risk management that are subsequent to the assessment stage. [V.C]

67. b; Supplier B has elevated scores in both key risk indicators, representing the highest total risk to the company. The others should continue to be monitored. [V.C]

68. d; Supplier D is rated high in both criticality and spend level, representing the highest level of risk to the company. The others should be monitored and periodically reevaluated. [V.C]

Section VI

Relationship Management

(47 questions)

A. SUPPLIER ONBOARDING

Understand and apply processes for orientation of suppliers such as providing overview of company, vision, mission, guiding principles, overall requirements, expectations, and criticality of product, service, and delivery requirements. (Apply)

B. COMMUNICATION

1. *Techniques and mediation*. Identify and apply communication techniques (e.g., oral, written, and presentation) specifically for internal stakeholders and suppliers to resolve issues. Apply different techniques when working in multicultural environments. Identify and describe the impact that culture, communications, and Diversity, Equity, and Inclusion (DEI) can have on an organization. (Evaluate)
2. *Reporting using quality tools*. Use appropriate technical and managerial reporting techniques for effective presentation and reporting, including the seven classic quality tools: Pareto charts, cause-and-effect diagrams, flowcharts, control charts, check sheets, scatter diagrams, and histograms. (Analyze)

C. LEADERSHIP AND COLLABORATION

Understand and apply techniques for coaching suppliers through regular communications, influencing without authority, negotiation techniques, conflict resolution techniques, and establish clear roles and responsibilities of internal stakeholders and suppliers using tools such as a RACI matrix (responsible, accountable, consulted, and informed). (Evaluate)

QUESTIONS

1. Which phase of the supplier orientation process is used to gather work requirements and methods, prepare work requirements for suppliers per scope of work, and establish expectations for suppliers?

   a. Prototyping

   b. Planning

   c. Development

   d. Execution

2. Which phase of the supplier orientation process is used to communicate the requirements, means and methods, schedules, and deliverables per scope of work?

   a. Prototyping

   b. Planning

   c. Development

   d. Execution

3. Which phase of the supplier orientation process is used to identify resources, processes, and work instructions to be shared with suppliers?

   a. Prototyping

   b. Planning

   c. Development

   d. Execution

4. One objective of supplier orientation is to:

   a. impart an understanding of the key work processes and requirements.

   b. introduce all the QA personnel and suppliers to each other.

   c. show the supplier your systems.

   d. give the supplier a chance to showcase their business to management.

5. One key outcome of supplier orientation is that:
   a. it creates happy suppliers.
   b. it builds the customer–supplier relationship.
   c. it lends visibility to expectations and the criticality of the outsourced work in the customer's eyes.
   d. the supplier and customer both know more about each other's businesses and processes.
6. Which of the following is a very important step in the orientation process?
   a. Cooperation
   b. Feedback
   c. Collaboration
   d. Good communication
7. This type of communicator is a good listener, avoids displays of power, and uses collaborative methods to find solutions to problems.
   a. Emotive
   b. Director
   c. Reflective
   d. Supportive
8. This type of communicator expresses themself in a more deliberate manner, may seem more quiet or introverted, and makes decisions more slowly.
   a. Emotive
   b. Director
   c. Reflective
   d. Supportive
9. A stern, serious attitude, strong opinions, and air of indifference are typical characteristics of this type of communicator.
   a. Emotive
   b. Director
   c. Reflective
   d. Supportive

10. While conducting global business transactions, cultural differences between the sender and receiver can have a significant bearing on the effectiveness of communication. Which of the following types of communication is most effective in this situation?

    a. E-mail

    b. Blog post

    c. Letter

    d. Presentation

11. A person from North America would be most likely to have which of the following communication styles?

    a. Low-context and sequential

    b. Low-context and synchronic

    c. High-context and sequential

    d. High-context and synchronic

12. A person from Asia would be most likely to have which of the following communication styles?

    a. Low-context and sequential

    b. Low-context and synchronic

    c. High-context and sequential

    d. High-context and synchronic

13. Which of the following can be described as the recognition of the different levels of knowledge, availability, and access available to different individuals and creating paths to equal opportunities for everyone?

    a. Inclusion

    b. Equality

    c. Diversity

    d. Equity

14. Which of the following can be described as the acknowledgement of the variation in the different people that one engages with regarding parameters such as gender, ethnicity, and socioeconomic status?
    a. Equality
    b. Diversity
    c. Inclusion
    d. Equity
15. Which of the following represents a business strategy to encourage purchasing from businesses that are owned and led by people from historically disadvantaged or underrepresented demographics?
    a. DEI
    b. Strategic sourcing
    c. Supplier diversity
    d. Supply chain rationalization
16. Which of the following types of cultures is characterized as having less explicit communication, less written and/or formal information, and more internalized understanding of what is being communicated?
    a. Synchronic
    b. High-context
    c. Affective
    d. Normative
17. In which of the following types of cultures are the past, present, and future intertwined when considering an issue?
    a. Normative
    b. Synchronic
    c. Sequential
    d. Affective

18. A process improvement team has completed a PFMEA. Which of the following tools would be best suited to help the team focus their efforts based on the PFMEA?

    a. SWOT analysis

    b. Pareto analysis

    c. Force-field analysis

    d. Statistical analysis

19. The phrase "vital few and useful many" is applicable to the:

    a. cause-and-effect diagram.

    b. check sheet.

    c. Pareto diagram.

    d. PFD.

20. Scatter diagrams are best described as which of the following?

    a. Correlation analysis

    b. Pareto analysis

    c. Process capability analysis

    d. Distribution analysis

Use the following figures to answer questions 21–24.

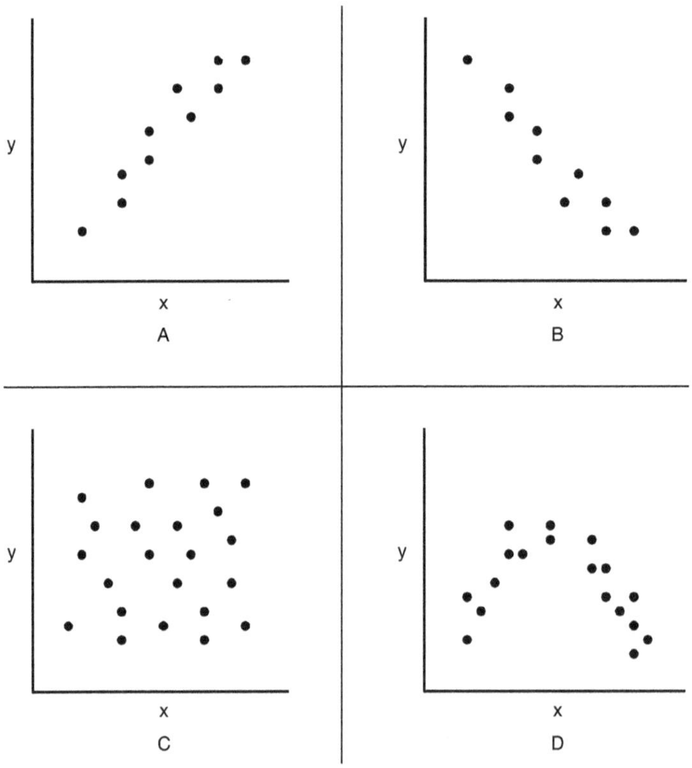

21. Which figure demonstrates negative correlation?
    a. Figure A
    b. Figure B
    c. Figure C
    d. Figure D
22. Which figure demonstrates nonlinear correlation?
    a. Figure A
    b. Figure B
    c. Figure C
    d. Figure D
23. Which figure demonstrates no correlation?
    a. Figure A
    b. Figure B
    c. Figure C
    d. Figure D
24. Which figure demonstrates positive correlation?
    a. Figure A
    b. Figure B
    c. Figure C
    d. Figure D
25. A tool used to provide visual identification of many potential causes of a problem is the:
    a. decision tree.
    b. process flowchart.
    c. cause-and-effect diagram.
    d. check sheet.

26. The tool that captures requirements on inputs into and outputs from a process is:
    a. requirements tree analysis.
    b. SIPOC.
    c. quality function deployment.
    d. the Kano model.
27. For the normal curve, approximately 68% of the data will fall between:
    a. $\pm 2\sigma$.
    b. $\pm 3\sigma$.
    c. $\pm 1\sigma$.
    d. $\pm 5\sigma$.
28. The tool used to monitor a process to make sure that the process improvements made are continuing to produce desired results is:
    a. a control chart.
    b. GR&R.
    c. a process map.
    d. a control plan.
29. Control charts are graphical tools used to monitor a process. Control charts generally do not contain or use:
    a. a centerline.
    b. control limits.
    c. specification limits.
    d. subgroups.
30. The quality tool that does not have a fixed form is the:
    a. control chart.
    b. scatter diagram.
    c. check sheet.
    d. cause-and-effect diagram.

31. Which quality tool is used to diagram a process?

    a. Control chart

    b. Scatter diagram

    c. Venn diagram

    d. Flowchart

32. Which of the following quality tools is best used to display the distribution of a set of data?

    a. Pareto chart

    b. Scatter diagram

    c. Control chart

    d. Histogram

33. Which of the following quality tools is best used to collect and tally data as it is encountered?

    a. Pareto chart

    b. Control chart

    c. Histogram

    d. Check sheet

34. Which of the following continuous improvement methods focuses on a business's customers, quality built into the work culture, and strong leadership involvement in quality efforts?

    a. Kaizen

    b. TQM

    c. PDCA

    d. Theory of constraints (TOC)

35. Armand V. Feigenbaum's concept of total quality includes:

    a. quality leadership, modern quality technology, and organizational commitment.

    b. commitment to quality, management commitment, and measurement of potential quality problems.

    c. statistical process control.

    d. benchmarking and reengineering.

36. In which role in the RACI model must the decision be discussed with the individual before a decision is made?

   a. Responsible

   b. Accountable

   c. Consulted

   d. Informed

37. In which role in the RACI model must the individual be cognizant about a decision because they are affected?

   a. Responsible

   b. Accountable

   c. Consulted

   d. Informed

38. In which role in the RACI model is the individual ultimately answerable for results?

   a. Responsible

   b. Accountable

   c. Consulted

   d. Informed

39. The type of leader who provides clear instruction and control to team members is:

   a. supportive.

   b. coaching.

   c. situational.

   d. directional.

Refer to the following chart to answer questions 40–41:

	Manager	**Engineer**	**Technician**	**Associate**
Activity 1			A	R
Activity 2			R	R
Activity 3	A	A	R	R
Activity 4	A	R	C	I

40. What is wrong with the line for Activity 2?

    a. There are two cells with an R.

    b. There is no cell with a C.

    c. There is no cell with an A.

    d. There is no cell with an I.

41. What is wrong with the line for Activity 3?

    a. There is no cell with a C.

    b. There are two cells with an R.

    c. There is no cell with an I.

    d. There are two cells with an A.

42. What type of leadership is best suited for high-performing team members that are technically competent?

    a. Delegating

    b. Directional

    c. Supportive

    d. Adversarial

43. Which of the following is the stage of negotiation where the parties work to understand each other's interests and develop mutually agreeable options?

    a. Discussion

    b. Resolution

    c. Planning

    d. Analysis

44. What are the four main points of strategy in negotiations?

    a. Analysis, Planning, Discussion, Resolution

    b. People, Interests, Options, Criteria

    c. Plan, Do, Check, Act

    d. People, Importance, Opportunity, Choice

45. Which of the following negotiation points focuses on separating the parties from the problem?

    a. People

    b. Interests

    c. Options

    d. Criteria

46. In which of the following negotiation points would the negotiator focus on benefits and not positions?

    a. Interests

    b. Options

    c. Criteria

    d. People

47. In which of the following negotiation points would negotiators focus on developing multiple alternatives to look for mutual gains before deciding what to do?

    a. Criteria

    b. Options

    c. Interests

    d. People

SOLUTIONS

1. c; The phase of the supplier orientation process used to gather work requirements and methods, prepare work requirements for suppliers per scope of work, and establish expectations for suppliers is the *development* phase. [VI.A]

2. d; The phase of the supplier orientation process used to communicate the requirements, means and methods, schedules, and deliverables per scope of work is the *execution* phase. [VI.A]

3. b; The phase of the supplier orientation process used to identify key resources, processes, work instructions, and methods to be shared with suppliers; set up communication with supplier contacts; and schedule orientation sessions is the *planning* phase. [VI.A]

4. a; Although (b), (c), and (d) are all important to accomplish during orientation, the real objective is for the supplier to understand processes and requirements. [VI.A]

5. c; (a) is desirable but not the answer to this question. (b) and (d) are also good answers, but as with (4) above, it's really important for the work and requirements to be understood from the customer point of view. [VI.A]

6. b; (a), (c), and (d) are all important attitudes and behaviors during the orientation process, but *feedback* is an actual step in the process. [VI.A]

7. d; A supportive communicator possesses a low degree of dominance and a high degree of sociability. This person is a good listener, avoids displays of power, and uses collaborative methods to find solutions to problems. An example of a supportive communicator is a Six Sigma Black Belt practitioner who leads open, low-stress meetings, encourages participation, and facilitates finding solutions that can be supported by all of those involved. [VI.B.1]

8. c; Someone who demonstrates low dominance and sociability is a reflective communicator. These individuals express themselves in a more deliberate manner, may seem more quiet or introverted, and make decisions more slowly. An example of a reflective communicator is a research scientist known for his or her lab work who prefers the ability to summarize and write white papers without dedicated due dates. [VI.B.1]

9. b; A director communicator ranks high on the dominance scale and low on the sociability scale. A stern, serious, indifferent attitude and strong opinions are typical characteristics of this type of communicator. An example of a director communicator is a company president who is very businesslike, expresses his or her opinions with conviction, and appears to be inflexible. [VI.B.1]

10. d; Avoiding cultural misunderstandings during written and presentation forms of communication is easier because the sender usually has much more time to develop the material before the delivery of the information. [VI.B.1]

11. a; A person from North America would be most likely to have a sequential and low-context communication style. [VI.B.1]

12. d; A person from Asia would be most likely to have a high-context and synchronic communication style. [VI.B.1]

13. d; This is a definition of equity. Equity and equality are different in that equity focuses on providing everyone with resources or opportunities specific to their needs, while equality focuses on giving everyone the same resources or opportunities regardless of their differences or what is actually needed. [VI.B.1]

14. b; This is a definition of diversity. [VI.B.1]

15. c; This is the purpose of a supplier diversity program. The terms in the other choices may include such a program but are too broad to be the best answer to this question. [VI.B.1]

16. b; High-context cultures tend to be less explicit in communication and depend more on the context (hence the name) and interpretation, while low-context cultures tend to be more explicit and detail-oriented in communication. [VI.B.1]

17. b; Synchronic cultures consider the past, present, and future together, believing that relationships are more important than strict schedules, while sequential cultures will observe schedules more closely and tend to address agenda items one-at-a-time. [VI.B.1]

18. b; A Pareto chart can be used to help a process improvement team focus their efforts based on the Process Failure Mode and Effects Analysis (PFMEA). [VI.B.2]

19. c; The purpose of the Pareto chart is to separate the *vital few* causes from the *trivial many*. This is often reflected in what is called the 80/20 rule, and helps focus attention on the more pressing issues. [VI.B.2]

20. a; Scatter diagrams visually display the strength and direction of the relationship between two variables. [VI.B.2]

21. b; This figure demonstrates negative correlation. The points fall on a straight line descending from left to right.

B

[VI.B.2]

22. d; This figure demonstrates nonlinear correlation (quadratic correlation). There are times when there can be linear, quadratic, cubic, or no correlation present.

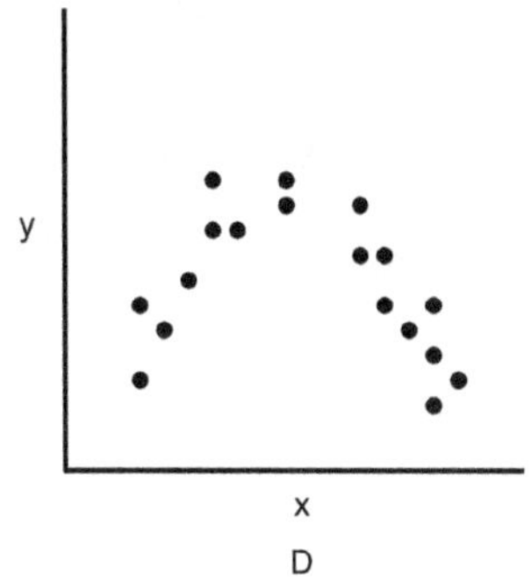

[VI.B.2]

23. c; This figure has no discernable relationship. Therefore, there is no correlation between the two variables.

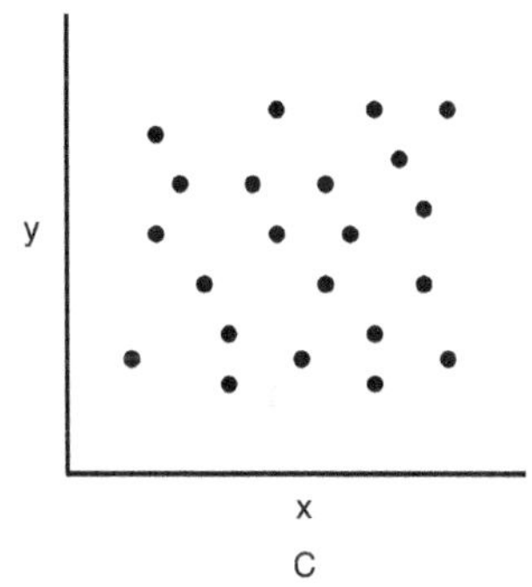

[VI.B.2]

24. a; This figure demonstrates positive correlation. The points fall on a straight line ascending from left to right. However, the correlation is not perfect due to the distance between the points (dispersion).

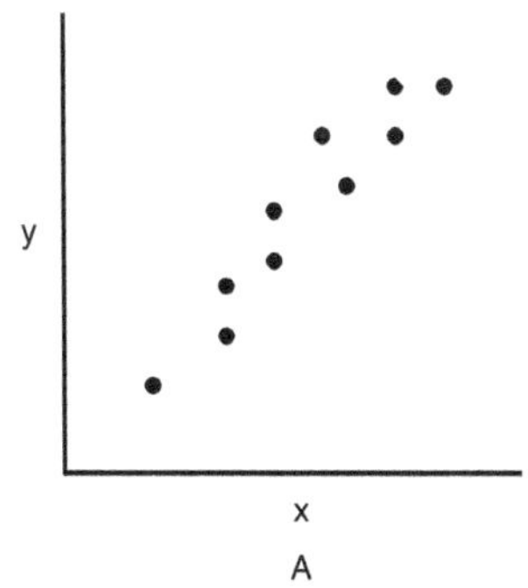

[VI.B.2]

25. c; A cause-and-effect diagram (also called an *Ishikawa diagram* or *fishbone diagram*) traditionally divides causes into several generic categories. In use, a large empty diagram is often drawn on a whiteboard or flip chart to visually display potential causes of a problem. [VI.B.2]

26. b; A key benefit of the SIPOC is that it is much easier to complete than either a process map or a value stream map. SIPOCs can be used as a basis for constructing detailed process maps and value stream maps at future dates. Further, SIPOCs help identify the voice of the customer as well as provide quick oversight into the initial *X*s and *Y*s. [VI.B.2]

27. c; The percentages for ±1, ±2, and ±3 standard deviations are provided in the figure. [VI.B.2]

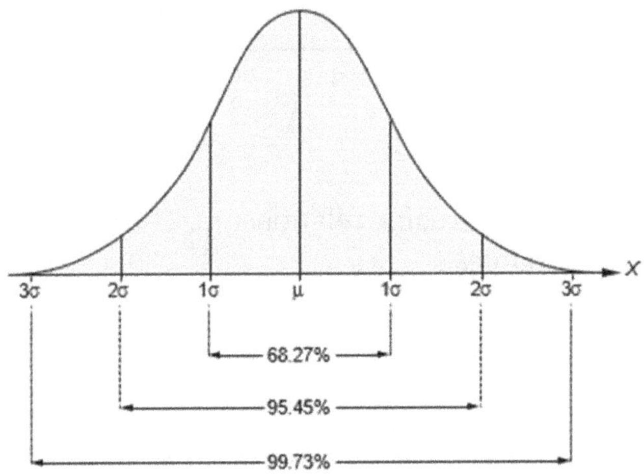

28. a; Once a process has been improved, it must be monitored to ensure that the gains are maintained and to determine when additional improvements are required. Control charts are used to monitor the stability of the process and determine when a special cause is present, and when to take appropriate action. [VI.B.2]

29. c; Control charts generally contain a centerline (process average) and upper and lower control limits (±3σ). Specification limits are generally not plotted or recorded on statistical control charts. [VI.B.2]

30. c; Check sheets are used to collect data in real time at the location where the data are generated. Check sheets can use quantitative or qualitative data. [VI.B.2]

31. d; Flowcharts are graphical representations of the steps in a process. Flowcharts are drawn to better understand processes. [VI.B.2]

32. d; Histograms are generally used to show and examine the distribution (shape) of a set of data. The other choices show and examine other aspects of a given data set. [VI.B.2]

33. d; The key phrase is "collect and tally," which is one of the main purposes of a check sheet. The other choices are options that may be used to display and evaluate data *after* it has been collected. [VI.B.2]

34. b; Total quality management (TQM) is an organization-wide quality program that focuses on the customers' involvement in determining quality, employees' obligation to quality, and strong leadership from upper management in quality improvement efforts. [VI.C]

35. a; Armand V. Feigenbaum lists three steps to quality: quality leadership, modern quality technology, and organizational commitment. [VI.C]

36. c; The RACI (responsible, accountable, consulted, and informed) matrix is usually defined in detail by a team or committee, and in some cases by an individual. The roles are:

    - *Responsible.* Individuals who actively participate in an activity.
    - *Accountable.* The individual ultimately accountable for results. Only one individual may be accountable at a time.
    - *Consulted.* Individuals who must be consulted before a decision is made.
    - *Informed.* Individuals who must be informed about a decision because they are affected. These individuals do not need to take part in the decision-making process.

    [VI.C]

37. d; The RACI (responsible, accountable, consulted, and informed) matrix is usually defined in detail by a team or committee, and in some cases by an individual. The roles are:

    - *Responsible.* Individuals who actively participate in an activity.
    - *Accountable.* The individual ultimately accountable for results. Only one individual may be accountable at a time.
    - *Consulted.* Individuals who must be consulted before a decision is made.
    - *Informed.* Individuals who must be informed about a decision because they are affected. These individuals do not need to take part in the decision-making process.

    [VI.C]

38. b; The RACI (responsible, accountable, consulted, and informed) matrix is usually defined in detail by a team or committee, and in some cases by an individual. The roles are:

    - *Responsible.* Individuals who actively participate in an activity.
    - *Accountable.* The individual ultimately accountable for results. Only one individual may be accountable at a time.
    - *Consulted.* Individuals who must be consulted before a decision is made.
    - *Informed.* Individuals who must be informed about a decision because they are affected. These individuals do not need to take part in the decision-making process.

    [VI.C]

39. d; Directional leadership is characterized by high directive and low supportive behavior. It is suited for individuals who possess low competence and high commitment. Such individuals are often described as the *enthusiastic beginner*. They are new to a task and motivated about it. [VI.C]

40. c; At minimum, each line in a RACI chart should have one, and only one, A (Accountable) role assigned. Without an A assigned, there is no one ultimately accountable for results. [VI.C]

41. d; Each line in a RACI chart should have one, and only one, A (Accountable) role assigned. If more than one role is assigned an A, it is unclear who is ultimately accountable for results. [VI.C]

42. a; Delegative leadership style is characterized by low directive and low supportive behavior. It is suited for individuals who possess high competence and high commitment. Such individuals are often described as the self-reliant. [VI.C]

43. a; The discussion phase of negotiations is where the parties should develop options that are beneficial to all involved. [VI.C]

44. b; The four main points of strategy to be used in all phases of negotiation are:

    - People: Separate the people from the problem.
    - Interests: Focus on interests, not positions.
    - Options: Develop multiple options that might be mutually beneficial before determining how to proceed.
    - Criteria: Require that results are based on a specified objective standard. [VI.C]

45. a; When focusing on people, the negotiator strives to separate the people from the problem. [VI.C]

46. a; When focusing on interests, the negotiator focuses on the parties' interests and not their positions. [VI.C]

47. b; When focusing on options, the negotiators develop multiple options that might be mutually beneficial before determining how to proceed. [VI.C]

Section VII

Business Governance, Ethics, and Compliance

(24 questions)

A. ASQ CODE OF ETHICS

Determine appropriate behavior in situations requiring ethical decisions, including identifying conflicts of interest, and recognizing and resolving ethical issues. (Apply)

B. COMPLIANCE AND SUSTAINABILITY

1. *Compliance.* Understand issues of compliance and their applicable policies, laws, and regulations (e.g., conflict of interest, confidentiality, and bribery). (Apply)
2. *Sustainability.* Understand and recognize the importance of environmental, social, and governance factors and adhere to applicable sustainability policies. (Understand)

C. CONFIDENTIALITY

1. *Organizational policies.* Apply organizational policies for executing appropriate agreements such as non-disclosure, quality, and change notification agreements. (Apply)
2. *Intellectual property.* Apply procedures for protecting the intellectual property of an organization and its suppliers. (Apply)
3. *Illegal activity.* Understand and interpret policies for reporting observations and deviations that could be perceived as illegal activity. (Apply)

QUESTIONS

1. A situation that exists when an individual exploits, or appears to exploit, their position for personal gain is a(n):

   a. emotional conflict.

   b. conflict of conscience.

   c. conflict of interest.

   d. conflict of effort.

2. The ASQ Code of Ethics requires ASQ members and ASQ certification holders during relations with the public to:

   a. perform services in any area of quality.

   b. hold paramount the safety, health, and welfare of the public in the performance of their professional duties.

   c. assure billing practices are fair.

   d. perform services for only one client at a time.

3. Supervisory independence and the use of non-executive board members is a:

   a. socially responsible investment (SRI).

   b. method to help minimize ethical issues.

   c. case for competitors as stakeholders.

   d. requirement of the ASQ Code of Ethics.

Use the following case study to answer questions 4–7.

You are performing a supplier qualification audit. The audit is not going well, but it's clear the supplier wants the business very badly, as does your R&D function. It's also clear the supplier can see it's going poorly. The general manager and quality assurance manager see gaps that your experience can fill. They ask you to help them to close those gaps. The gaps are serious but not critical or safety related. They are not a sole supplier or sole source.

4. Is it okay to help with those gaps?

   a. No. An auditor must remain objective. After the audit, if all parties see value in you helping, it could be acceptable.

   b. Yes, because they are trying their best and have not offered money or any incentive.

   c. Yes, because R&D will say supplier quality is setting up obstacles again.

   d. No, because it makes the auditor liable for the results.

5. How should you handle this situation?
   a. End the audit. This supplier is not qualified and will fail the audit anyway.
   b. Explain that you are flattered but you must remain objective, and continue.
   c. Help them set up an action plan as you go forward so they have each finding covered.
   d. Ignore the request as addressing it will only increase the awkwardness.
6. At the conclusion of the audit, you should tell your manager:
   a. about the incident, how you handled it, and why.
   b. nothing. It is not a big deal, you did the right thing. It'll call your skills into question.
   c. how unqualified they are. End the issue by not qualifying the supplier.
   d. you'll try to institute damage control with R&D, or they will blame you for not qualifying their supplier.
7. How would this scenario change if the supplier *did* offer you pay for consulting?
   a. It doesn't. Handle it the same way.
   b. It now becomes a higher level of conflict of interest because they are trying to incentivize you.
   c. It's still okay as long as you consult on your free time.
   d. It doesn't. It's not a bribe to alter the audit report or anything dishonest.
8. Which of the following refers to the adherence to laws, regulations, and standards that are set by government agencies, industry associations, and other relevant organizations?
   a. Social responsibility
   b. Compliance
   c. Sustainability
   d. Ethics
9. Which of the following is provided by a supplier of a product subject to the FD&C Act as assurance that the product is not adulterated or misbranded?
   a. Warranty
   b. Guaranty
   c. Surety
   d. Bond

10. Which of the following are FDA regulations that summarize the correct procedures in food production or processing facilities?

    a. GLPs

    b. HACCP

    c. GMPs

    d. REACH

11. Which of the following is a system that identifies and monitors safety criteria in a process to prevent hazards from occurring?

    a. REACH

    b. HACCP

    c. RoHS

    d. GLPs

12. Each of the following is a main pillar of sustainability *except*:

    a. economic sustainability.

    b. social sustainability.

    c. environmental sustainability.

    d. governance sustainability.

13. Which of the following refers to the practices and processes that allow an organization to operate in a manner that meets the needs of the present generation without compromising the ability of future generations to meet their own needs?

    a. Social responsibility

    b. Sustainability

    c. Compliance

    d. Strategic planning

14. The reduction of waste and emissions and use of renewable energy sources are examples of which main pillar of sustainability?

    a. Environmental

    b. Economic

    c. Social

    d. Strategic

15. Efforts to reduce costs and to invest in technology and infrastructure is an example of which main pillar of sustainability?
    a. Economic
    b. Social
    c. Environmental
    d. Strategic
16. Supplier quality agreements should include all of the following *except*:
    a. a change notification process.
    b. a nondisclosure agreement/IP protections.
    c. applicable quality system requirements.
    d. purchasing terms.
17. A statement of work (SOW):
    a. is most appropriate for a service provider (lab, consultant).
    b. can be used interchangeably for service and manufacturing suppliers.
    c. provides details on expectations and deliverables.
    d. (a) and (c).
18. Purchasing terms:
    a. are appropriate to include in a quality agreement.
    b. should be kept apart from a quality agreement.
    c. are handled by supplier quality professionals.
    d. dictate the type of quality agreement.
19. In the United States, IP violations may be reported to:
    a. the US Patent Office.
    b. the US Attorney's Office.
    c. the National Intellectual Property Rights Coordination Center.
    d. your member of Congress.

20. WIPO is an organization that:
    a. protects IP rights globally by administering treaties.
    b. polices violations of IP.
    c. offers registration of IP.
    d. coordinates with the US Patent Office.
21. IP infringements should be handled __________ with the supply base.
    a. on a case-by-case basis
    b. only as needed
    c. within a standard, documented process
    d. swiftly and punitively
22. The difference between illegal activities and conflicts of interest is:
    a. semantics. The real difference is in the quantity of incentives offered.
    b. that laws vary from jurisdiction to jurisdiction, so it's a judgement call.
    c. that someone can go to prison for one and not the other.
    d. that one is a clear violation of a legal statute and the other is an ethical violation.
23. Illegal activities include all but (choose one):
    a. paying for a private practitioner's airfare.
    b. bribes to officials.
    c. trading in illegal materials.
    d. trading in restricted materials (certain types of technology).
24. US law provides protection to *whistleblowers* under the:
    a. Taft–Hartley Amendments.
    b. Sarbanes–Oxley Act.
    c. Dodd–Frank Act.
    d. Wagner Act.

SOLUTIONS

1. c; A conflict of interest occurs when an individual acts in a way that exploits a given situation to better themselves. [VII.A]

2. b; The ASQ Code of Ethics requires ASQ members and ASQ certification holders in Article 1 to hold paramount the safety, health, and welfare of the public in the performance of their professional duties in relation with the public. [VII.A]

3. b; Supervisory independence and the use of non–executive board members are methods to help minimize ethical issues. [VII.A]

4. a; This is a conflict of interest. Objectivity is the auditor's currency. However, since they might become a business partner if all parties agree and the audit results stand objectively, it could be acceptable after the audit concluded. The other answers do not demonstrate an understanding of ethics and conflicts of interest. [VII.B.1]

5. b; There is no need to end the audit at this point. Remain objective and polite but carry on. The other answers do not demonstrate an understanding of ethics and conflicts of interest. [VII.B.1]

6. a; You must report the offer—it could be indicative of future issues. Hiding it and burying it won't help, and could be perceived as dishonest. The other answers do not demonstrate an understanding of ethics and conflicts of interest. [VII.B.1]

7. b; The situation is now serious, and crossed any line of possible misunderstanding the minute money was offered. The other answers do not demonstrate an understanding of ethics and conflicts of interest. [VII.B.1]

8. b; This is a definition of compliance. [VII.B.1]

9. b; The Food, Drug, and Cosmetic (FD&C) Act requires a guaranty from the supplier. [VII.B.1]

10. c; This is a definition of Good Manufacturing Practices (GMPs). Good Laboratory Practices (GLPs) are similar but cover lab practices, while Registration, Evaluation, Authorization, and Restriction of Chemicals (REACH) covers the control of hazardous chemicals/materials. The use of Hazard Analysis Critical Control Point (HACCP) is ultimately voluntary in the United States, although it may be required by specific customers. [VII.B.1]

11. b; HACCP is used in multiple industries to monitor safety criteria in a process for the prevention of hazards. GLPs cover best practices in a lab environment, while REACH and Restriction of Hazardous Substances (RoHS) cover control of dangerous materials but not the processes in which they are generated or introduced. [VII.B.1]

12. d; Governance is not a main pillar of sustainability, although good governance may be needed to support the pillars. [VII.B.2]
13. b; This is a definition of sustainability. Social responsibility is a broader topic. [VII.B.2]
14. a; The references to emissions and renewable energy indicate that these fall under the environmental pillar. [VII.B.2]
15. a; Cost reductions and investments in technology and infrastructure are key examples within the economic pillar. [VII.B.2]
16. d; (a), (b), and (c) are typically found in a quality agreement. Purchasing terms should be included in a contract, not in a quality agreement. [VII.C.1]
17. d; Although it's not impossible to apply a Statement of Work (SOW) to a manufacturing supplier, it would not be the best format choice. (a) and (c) are found in typical SOWs. [VII.C.1]
18. b; Purchasing terms should be kept apart from a quality agreement. [VII.C.1]
19. c; US Immigration and Customs Enforcement (ICE) runs this office and any enforcement will flow from reporting to the national office. The US Attorney might enforce after reporting. The Patent Office and Congress provide the means and the laws for registration of IP. [VII.C.2]
20. a; World Intellectual Property Organization (WIPO) is an organization that administers global treaties. These other activities are for individual nations to carry out. [VII.C.2]
21. c; Intellectual Property (IP) needs to be handled up front and documented with suppliers in a uniform way to prevent issues later. [VII.C.2]
22. d; (c) is true but not the most accurate distinction. (a) and (b) are false. [VII.C.3]
23. a; (b), (c), and (d) are all clearly illegal. [VII.C.3]
24. c; Whistleblowers are protected under the Dodd–Frank Act. (d) is a labor law dealing largely with unions and overtime, (a) is an amendment to the Wagner Act, and the Sarbanes–Oxley Act is a finance law enacted in the aftermath of the Enron scandal. [VII.C.3]

Part B
Practice Exam

(150 questions)

QUESTIONS

1. A supplier quality engineer is managing risk for a supplier that produces critical components for a defense contract. The supplier has been identified as a single-source supplier, which increases the risk of supply chain disruptions. What is the most effective action the engineer should take to control and minimize this risk?

    a. Increase safety stock levels of the critical components.

    b. Establish a more frequent audit schedule for the supplier.

    c. Develop an emergency response plan and identify alternative suppliers.

    d. Negotiate longer lead times to give the supplier more production flexibility.

2. A supplier quality engineer is working with the purchasing department to finalize the terms and conditions (T&C) of a new supplier contract for custom-made components. Which of the following clauses should be included in the T&C to ensure the supplier is liable for delivering products that meet the specified quality and performance standards?

    a. A clause that defines the supplier's warranty coverage for delivery delays

    b. A clause outlining the procedure for managing engineering change requests

    c. A clause specifying penalties for nonconformance to quality standards

    d. A clause requiring the supplier to follow general industry best practices

3. Audit reports should include all of the following *except*:
   a. argumentative statements.
   b. subjective opinions.
   c. proprietary information.
   d. details of observed evidence.
4. Implementing the solution is associated with which phase of the PDCA cycle?
   a. Plan
   b. Do
   c. Check
   d. Act
5. Comparison of an actual operation to the documented requirements of that operation is an example of which of the following types of audits?
   a. Process audit
   b. Product audit
   c. Material audit
   d. Management system audit
6. Increased globalization and outsourcing have given rise to the introduction of what type of materials and parts within the supply chain?
   a. Counterfeit
   b. Unique
   c. Complex
   d. Complete
7. A customer order for 1500 parts is due in three days. ABC runs one eight-hour shift per day. What is the takt time (in minutes per unit)?
   a. 0.90
   b. 0.96
   c. 1.50
   d. 1.60

8. Requirements regarding FAI and part qualification procedures should be outlined in which of the following?

   a. Quality agreement

   b. Product specifications

   c. Engineering drawings

   d. Inspection plans

9. Which of the following is defined as a situation in which critical information is withheld from the team because individual members censor or restrain themselves, either because they believe their concerns are not worth discussing or because they are afraid of confrontation?

   a. Storming

   b. Socializing

   c. Brainstorming

   d. Groupthink

10. A critical-to-quality (CTQ) characteristic is usually determined by:

    a. the effect on form/fit/function.

    b. customer perception of the overall quality of that characteristic.

    c. the need to inspect this characteristic.

    d. (a) and (c).

11. Which of the following can be described as a frequency distribution graph showing the number of observations made of a particular parameter?

    a. Histogram

    b. Scatter diagram

    c. Box plot

    d. Run chart

12. Why is it important to collaborate with the supplier and get supplier feedback on requirements?

    a. Suppliers should listen to their customers and follow the contract.

    b. The supplier needs to own adherence to the specifications once they are agreed to.

    c. The supplier is an expert in the process they perform and has a great deal of experience translating customer requirements into their processes.

    d. The supplier most likely has more process knowledge about their process.

13. The implementation of fair labor business practices is an example of which main pillar of sustainability?

    a. Strategic

    b. Social

    c. Economic

    d. Environmental

14. A high-performing team has recently had changes to the individuals participating in the team project. With the change in personnel, in what stage is the team likely to find itself?

    a. Performing

    b. Norming

    c. Storming

    d. Forming

15. Effective joint quality planning between a customer and supplier involves detailed discussions and review of each of these areas *except*:

    a. cultural.

    b. economic.

    c. managerial.

    d. technological.

16. Commitment and trust are elements of:

    a. partnering.

    b. sourcing.

    c. development.

    d. improvement.

17. Defects should be classified according to:

    a. acceptability.

    b. quantity.

    c. severity.

    d. size.

18. A company rates its raw material suppliers based on two equally important key risk factors, nonconformance rate and incoming inspection reject rate. The ratings for the past quarter are included in the following table. Higher values indicate higher levels of risk. Based on this information, which of these suppliers represents the largest risk to the company?

Supplier	Nonconformance Rate	Incoming Inspection Rate
A	3	1
B	3	5
C	5	5
D	3	3

    a. A

    b. B

    c. C

    d. D

19. A supplier quality engineer is collaborating with the purchasing department during a supplier selection process. Beyond pricing, which of the following tasks would the purchasing department typically handle during supplier qualification?

    a. Defining product specifications and performance criteria

    b. Negotiating delivery schedules and payment terms

    c. Developing supplier corrective action plans for defects

    d. Conducting a detailed analysis of the supplier's quality system

20. Which of the following can be used to evaluate potential cause-and-effect relationships?

    a. Run charts

    b. Control charts

    c. Scatter diagrams

    d. Histograms

21. Which of the following can be defined as the time between initiation of a project and its completion?
    a. Takt time
    b. Slack time
    c. Lead time
    d. Cycle time
22. When a company audits its own processes, it is conducting which type of audit?
    a. First party
    b. Second party
    c. Third party
    d. Independent
23. The objective of finalizing requirements collaboratively is:
    a. to simultaneously meet goals for quality, cost, and delivery.
    b. to make sure all parties agree.
    c. so the supplier can't say later, "I didn't know."
    d. to make sure purchasing meets their spend reduction targets.
24. Which of the following shows the relationship between requirements, risks, and needs of the customer?
    a. Demand plan
    b. House of quality
    c. PFMEA
    d. Design output
25. Which of the following could be used to help identify all of the internal and external customers of a process?
    a. FMEA
    b. Pareto chart
    c. FTA
    d. PFD

26. Available data do not always provide a reliable basis for predicting the future. For unique types of risks or for new product types, _____ data may not be available.

    a. attribute

    b. variables

    c. neutral

    d. historical

27. A facilitator can turn team conflict into a problem-solving process by:

    a. discussing conflict resolution techniques.

    b. developing rules for conflict resolution.

    c. identifying the root cause analysis.

    d. evaluating concerns.

28. Which of the following can be described as a set of regulations regarding the elimination or minimization of potential issues of contamination or other processing errors in a manufacturing environment through the proactive implementation of sanitation, cleanliness, equipment verification, and process validation?

    a. GLP

    b. GMP

    c. REACH

    d. ITAR

29. Each of the following situations represents a potential conflict of interest *except*:

    a. An auditor is not assigned to audit a process within their own department.

    b. A consultant has contracts to work with two competing firms simultaneously.

    c. A company chooses a supplier that is owned by the purchasing manager's cousin.

    d. A person advocates for choosing a supplier in order to receive a signing bonus from that supplier.

30. Capability studies (assuming using Cpk) show all of the following *except*:

    a. how many parts are predicted to be out of spec, based on the Cpk result.

    b. how centered the distribution is within spec limits.

    c. whether parts are functional and mate with other components well.

    d. the relationship of the process spread to the specification limits.

31. A supplier quality engineer is working with the design team to address a known hazard in a product that cannot be eliminated through design changes. What should the engineer recommend incorporating into the product to mitigate the risk and ensure user safety?

    a. Safety interlocks and alarms

    b. Disclaimers and interlocks

    c. Disclaimers and safety manuals

    d. Alarms and disclaimers

32. Which of the following can be described as a management system in which food safety is addressed through the analysis and control of biological, chemical, and physical hazards throughout the manufacturing process, from raw material procurement through consumption?

    a. GLP

    b. REACH

    c. RoHS

    d. HACCP

33. In which of the following types of cultures are people more likely to show emotion and be more tolerant of emotional displays in a communication session?

    a. Synchronic

    b. Normative

    c. High-context

    d. Affective

34. When determining a make/buy decision, which tool would be best for analyzing historical performance?

    a. DMAIC

    b. PDCA

    c. CAPA

    d. FMEA

35. A process improvement team has completed a PFMEA. Which of the following tools would be best suited to help the team focus their efforts based on the PFMEA?

    a. SWOT analysis

    b. Pareto analysis

    c. Force-field analysis

    d. Statistical analysis

36. Audit records include the following *except*:

    a. auditor training and credentials.

    b. auditors' notes.

    c. the audit plan.

    d. CAPA records.

37. Which of the following categories describes a supplier that has consistently produced quality better than minimum requirements and may receive bidding priority?

    a. Approved

    b. Preferred

    c. Certified

    d. Qualified

38. Someone from which of the following cultural styles would benefit from a greater amount of detail when communicating specifications, nonconformances, and metrics?

    a. Low-context

    b. High-context

    c. Sequential

    d. Synchronic

39. Following are some of the steps of the 8D method. Put them in the correct order.

    1. Choose and verify permanent corrections.
    2. Determine, identify, and verify root causes and escape points.
    3. Implement and validate corrective actions.
    4. Develop containment plan and implement/verify interim actions.

    a. 4, 1, 3, 2
    b. 4, 2, 1, 3
    c. 2, 4, 1, 3
    d. 2, 1, 3, 4

40. Regarding a measurement system, the target below represents which of the following?

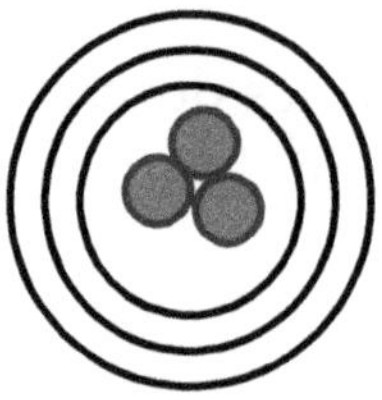

    a. Good accuracy and good precision
    b. Poor accuracy and good precision
    c. Good accuracy and poor precision
    d. Poor accuracy and poor precision

41. Which of the following refers to the path through the project plan on which all activities have zero slack time?

    a. Priority path
    b. Strategic path
    c. Key path
    d. Critical path

42. The EU's REACH and RoHS are examples of:

    a. hazardous materials regulations.
    b. medical device regulations.
    c. food safety regulations.
    d. employment laws.

43. Heijunka boxes or boards are typically implemented to help address which of the following wastes?
    a. Inventory
    b. Excessive processing
    c. Transportation
    d. Defects

44. The main purpose of a ____________ is to detect assignable causes of variation in a process.
    a. run chart
    b. scatter diagram
    c. Pareto chart
    d. control chart

Use the following information to answer questions 45–46.

Number of orders = 58

Number of orders delivered on time = 56

Number of orders delivered in full on time = 52

45. What is the OTD percentage?
    a. 76.0%
    b. 89.7%
    c. 96.6%
    d. 97.5%

46. What is the OTIF percentage?
    a. 76.0%
    b. 89.7%
    c. 96.6%
    d. 97.5%

47. All of the following are elements of PPAP *except*:

    a. FMEA.

    b. supplier quality audit.

    c. control plan.

    d. initial process studies.

48. During an FAI of a newly developed aerospace component, a supplier quality engineer discovers that a critical dimension exceeds its tolerance. Which of the following actions should the engineer take to ensure the FAI is properly completed?

    a. Adjust the tooling and restart production after rechecking the dimension.

    b. Recalibrate the measurement equipment and proceed with the FAI.

    c. Investigate the cause of the dimensional error, correct it, and conduct a new FAI.

    d. Continue production but add additional inspections for this dimension.

49. Place the following steps for developing standard work in order.

    I. Determine the amount of time required to perform the task.

    II. Determine the process and draft the initial document.

    III. Determine the scope of work.

    IV. Determine what tools are required.

    a. III, II, IV, I

    b. III, IV, I, II

    c. IV, III, II, I

    d. IV, II, I, III

50. A three-factor full factorial experiment was conducted. Each factor has two levels. Based on the following interaction plot, which is the most significant interaction?

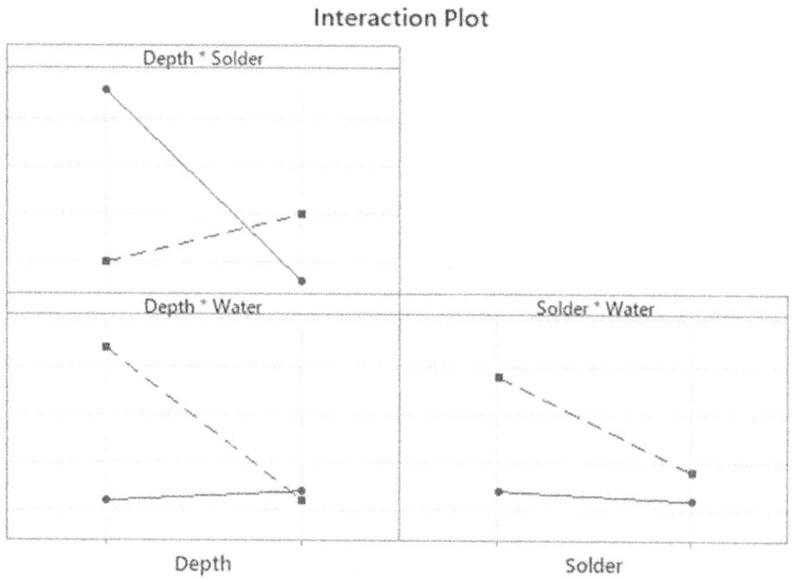

a. Depth * Solder

b. Depth * Water

c. Solder * Water

d. None of the interactions are significant.

51. Reporting of each of the following would be considered within the scope of whistleblower protection *except*:

a. violation of a federal law.

b. abuse of authority.

c. a policy dispute between the person reporting and a manager.

d. gross financial mismanagement.

52. This team role assists the lead auditor in organizing the team and making it more effective, but does not participate as a team member.

a. Client

b. Auditor

c. Facilitator

d. Lead auditor

53. A supplier quality engineer is monitoring the performance of a critical supplier that produces components for a high-risk medical device. The supplier has experienced several minor quality issues over the past six months. Which of the following actions should the engineer prioritize to continuously monitor and mitigate potential risks?

    a. Implement real-time monitoring of key quality metrics and establish trigger points for action.

    b. Schedule quarterly performance review meetings to address any ongoing issues.

    c. Request monthly updates on the supplier's production schedule and financial health.

    d. Conduct a one-time comprehensive audit of the supplier's entire quality management system.

54. A recall due to an airbag malfunction on an automobile was issued by a car company. How would the company classify this type of defect?

    a. Minor defect

    b. Major defect

    c. Serious defect

    d. Critical defect

55. What to do when the critical parameter on the control plan goes out of control is indicated on the control plan. It would most likely be found in the ___ column.

    a. reaction plan

    b. characteristic

    c. specification

    d. frequency

56. Empowering process improvement team members to function independently is known as _________ leadership.

    a. situational

    b. supportive

    c. deductive

    d. directional

57. The type of risk involved when a lot could be accepted when it does not conform to the specification is known as:

    a. Ω.

    b. α.

    c. β.

    d. σ.

58. __________ may include qualified auditors, auditors in training, and technical experts.

    a. Clients

    b. Audit team members

    c. Facilitators

    d. Lead auditors

59. Large conveyor systems and fleets of forklifts can indicate which of the following wastes?

    a. Waiting

    b. Excessive processing

    c. Inventory

    d. Transportation

60. Which of the following translates customer expectations for a product's function into specific engineering and quality characteristics?

    a. QFD

    b. Process development

    c. Design review

    d. FMEA

61. Intellectual property can be protected through each of the following mechanisms *except*:

    a. patents.

    b. trade secrets.

    c. trademarks.

    d. copyrights.

62. Incoming acceptance can be done through either lot or part acceptance. Apex Manufacturing, a supplier of key components, shipped 35 lots in Q1, of which two were rejected. What is the percentage of lot acceptance for Q1?

    a. 5.71%

    b. 6.06%

    c. 94.29%

    d. 100.00%

63. What is the primary difference between vision and mission statements?

    a. Mission statements relate to the future of the company.

    b. Vision statements describe the company's purpose.

    c. Mission statements explain the company's reason for existence.

    d. Mission and vision statements are essentially the same and are interchangeable.

64. Which of the following is represented by the symbol ||?

    a. Straightness

    b. Parallelism

    c. Flatness

    d. Perpendicularity

65. The ASQ Code of Ethics applies to:

    a. ASQ members.

    b. ASQ certification holders.

    c. (a) and (b).

    d. all quality professionals.

66. An individual who is an easygoing, down-to-earth quality manager who leads by example, has a friendly demeanor, and gets people to follow them without difficulty is what type of communicator?

    a. Emotive

    b. Director

    c. Reflective

    d. Supportive

67. Intellectual property can be protected:

    a. by lawsuits and court actions.

    b. only in the United States.

    c. by registering with the WIPO Convention.

    d. by applying for patents, copyrights, and trademarks.

68. Kanban uses the ________________ to determine the appropriate amount of inventory.

    a. lead time

    b. wait interval

    c. on-time delivery metric

    d. production consumption rate

69. Responsibilities of the supplier during the supplier orientation process include each of the following *except*:

    a. gain an understanding of work processes and requirements.

    b. define the frequency of required communications.

    c. clarify any issues, discrepancies, or ambiguities regarding the work requirements.

    d. understand final customer expectations of the supplied product.

70. Deviations from specifications might be acceptable if:

    a. engineering signs off on use as-is status.

    b. procurement gets the supplier to lower their piece part price sufficiently to pay for more incoming inspection.

    c. there is sufficient engineering/scientific objective evidence there will be no adverse effect on form/fit/function, performance, or safety.

    d. there is sufficient need to launch the product on time and within budget.

71. Which of the following can be used to compare product from several suppliers to the same specification?

    a. Pareto chart

    b. Scatter diagram

    c. Run chart

    d. Histogram

72. Which tool would be the most useful for analyzing categories of failures?
    a. Cause-and-effect charts
    b. Pareto charts
    c. Data charts
    d. Control charts
73. An effective supplier management program can reduce the risk involved when there are routine ________ with the supplier.
    a. communications
    b. mitigation controls
    c. purchase orders
    d. inspections
74. Which of the following would be the most appropriate tool to directly gauge the knowledge of students in a training class and determine which topics may need additional emphasis?
    a. Survey
    b. Pre-training test
    c. Multiple-choice test
    d. Interviews
75. The "C" in DMAIC is critical. Why?
    a. It stands for cost and serves as justification for the entire project.
    b. It stands for control, specifically of costs and employee behavior.
    c. It stands for control, and is vital to maintain the gains achieved by the project.
    d. It stands for check and is vital to determine the effects of the implemented actions.
76. In which of the following negotiation points would the negotiator insist that the results be based on some objective standard?
    a. Interests
    b. People
    c. Criteria
    d. Options

77. Which perspective defines an organization's value proposition and measures how effective the organization is in creating value?

a. Internal business processes perspective

b. Learning and growth perspective

c. Financial perspective

d. Customer perspective

78. Based on the following risk matrix table stratifying a company's suppliers, which supplier represents the lowest level of risk to the company?

	Parts Criticality—Low	Parts Criticality—High
Spend Level—Low	A	B
Spend Level—High	C	D

a. A

b. B

c. C

d. D

79. ____________ is/are a key aspect of supplier orientation.

a. Customer satisfaction

b. Piece part price agreement

c. Documentation of customer–supplier disagreements

d. Clearly defined roles and responsibilities

80. The function of the ____________ is to review nonconforming material and determine its disposition.

a. engineering department

b. MRB

c. quality engineer

d. supplier representative

81. Movement beyond a transactional customer–supplier relationship and toward actual coordination of functional activities is a hallmark of which of the following?

    a. Supplier selection

    b. Supplier qualification

    c. Supplier certification

    d. Supplier partnership

82. Which of the following tools would be the most appropriate for crisis management?

    a. PEST analysis

    b. SWOT analysis

    c. Contingency planning

    d. Risk analysis

83. Which of the following tools is best used when defining the scope of a process audit?

    a. Fishbone diagram

    b. SIPOC diagram

    c. Pareto chart

    d. Control plan

84. Which of the following is an example of a quality management specification?

    a. ISO 9001

    b. ITAR

    c. REACH

    d. RoHS

85. Which of the following is best described as the process of integrating a new supplier into a company's business processes, requirements, and communication practices?

    a. Classifying

    b. Selecting

    c. Onboarding

    d. Outsourcing

86. How is the audit team leader usually determined?

a. Chosen by audit team members

b. Selected by the client or audit program manager

c. Given to the auditor with the most audit experience

d. Given to the highest-ranked auditor on the team

87. Percent on-time delivery and percent late deliveries are examples of:

a. timeliness metrics.

b. quality metrics.

c. delivery (JIT) metrics.

d. compliance metrics.

88. Supplier remediation usually occurs:

a. at the qualification phase.

b. after performance issues arise.

c. as a result of the supplier's internal audit.

d. after a regulatory inspection.

Use the following risk table to answer questions 89–90. Higher values are less desirable.

Supplier	QMS maturity	Weight	Value	Financial	Weight	Value	Capacity	Weight	Value	Totals
A	3	3	9	3	1	3	3	2	6	162
B	2	3	6	4	1	4	1	2	2	48
C	1	3	3	2	1	2	2	2	4	24
D	4	3	12	1	1	1	3	2	6	72

89. ________ is the most critical category.

a. QMS maturity

b. Capacity

c. Finance

d. All are equal.

90. Which is the lowest-risk supplier?

    a. A

    b. B

    c. C

    d. D

91. Which of the following can be used to review and understand the flow of a process?

    a. FTA

    b. Control chart

    c. VSM

    d. Cause-and-effect diagram

92. In a lean system, what is another name for waste?

    a. Kanban

    b. Kaizen

    c. Seiton

    d. Muda

93. In a project plan, which of the following is best defined as an indication of how much delay can be tolerated for an event or task before the entire project is delayed?

    a. Excess time

    b. Lead time

    c. Slack time

    d. Down time

94. Each of the following should be addressed in a quality agreement with a supplier *except*:

    a. the supplier's responsibilities regarding receipt and storage of required materials.

    b. nondisclosure of proprietary information and processes.

    c. management of intellectual property.

    d. timely notification of process and/or product changes.

95. A supplier quality engineer is reviewing a new supplier agreement for a critical component. The contract includes terms for pricing, delivery schedules, and performance metrics, but does not specify how to handle recurring quality defects. What clause should be added to ensure the supplier is responsible for identifying and fixing the source of these defects?

a. A liquidated damages clause to impose penalties for late deliveries

b. A rework and repair clause to outline compensation for defective products

c. A continuous improvement clause to encourage long-term quality enhancements

d. A clause that defines how the supplier must address and resolve quality failures

96. Which of the FTA symbols indicates an undeveloped event?

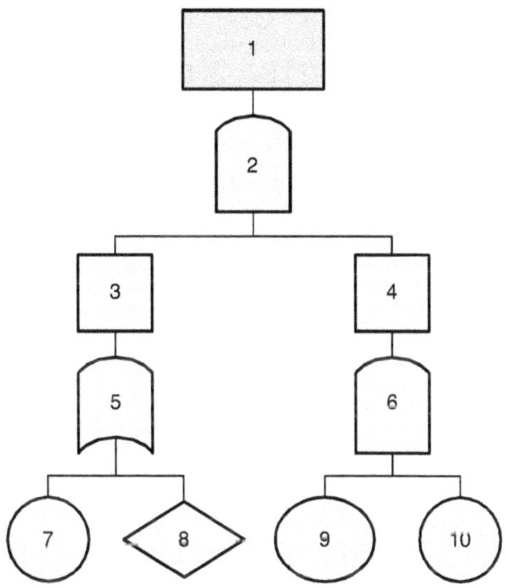

a. 7

b. 8

c. 9

d. 10

Use the following table to answer questions 97–98:

	Original process				Action results			
Row	Severity	Occurrence	Detection	RPN	Severity	Occurrence	Detection	RPN
1	3	5	5	75	3	5	3	45
2	5	10	3	150	3	3	5	45
3	10	3	5	150	5	3	3	45

97. On which row of the FMEA should a process improvement team focus their efforts for the original process state?

    a. Row 1

    b. Row 2

    c. Row 3

    d. Since RPNs are the same, no special effort is necessary.

98. Rank the order of risk for the process after mitigation actions have been completed.

    a. 3, 1, 2

    b. 2, 3, 1

    c. 1, 2, 3

    d. Since RPNs are the same, the risk is the same.

99. The change in bias over time when using a measurement system to measure a given part is referred to as measurement system ____________.

    a. accuracy

    b. stability

    c. repeatability

    d. reproducibility

100. Which of the following can be described as executing plans, conducting communications, and doing business in ways that include everyone in the organization?

    a. Diversity

    b. Equality

    c. Inclusion

    d. Equity

101. A supplier quality engineer has received data from the executed qualification plan for a new medical device component. The data includes measurements of CTQ characteristics, results from the FAI, and the supplier's CoA. Which of the following actions should the engineer take when collaborating with the internal team to interpret these results?

   a. Validate the supplier's production process results against the approved control plans.

   b. Focus only on reviewing the CTQ data to assess the product's overall quality.

   c. Recommend reducing in-process inspections to improve production speed.

   d. Accept the supplier's CoA without further review.

102. Handling the supplier qualification audit well, including open communication, collaboration, and meeting commitments, is important because:

   a. compliance is a critical requirement for suppliers to establish and maintain.

   b. it might well be a first impression for both parties and therefore critical to the future relationship.

   c. it is a necessity to begin the process of onboarding suppliers.

   d. it can give insights into the supplier's systems and controls that a questionnaire can't.

103. The registration adopted to meet the EU requirements to protect human health and the environment, and to prevent risk posed by chemicals, is known as:

   a. REACH.

   b. RoHS.

   c. OSHA.

   d. Conflict minerals.

104. How may out-of-control conditions exist in this control chart?

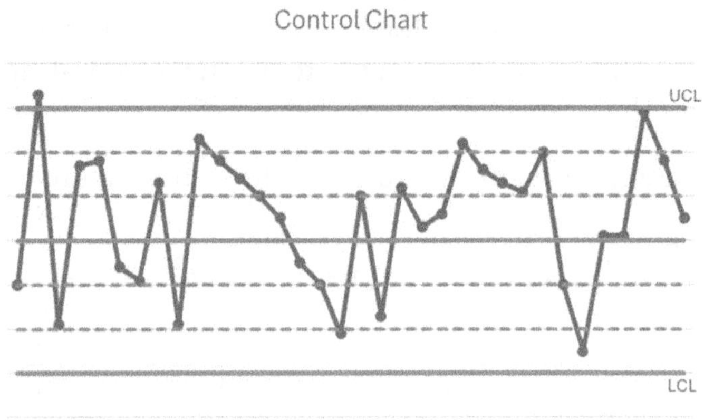

a. 1

b. 2

c. 3

d. 4

105. Nonconforming material has been previously identified and segregated, but the production manager believes that the nature of the nonconformance is small enough that the material can be used. What should happen next?

a. The material should be released.

b. The MRB should review the material for concurrence.

c. The company should follow its formal, documented procedure.

d. The MRB should obtain approval from engineering for use of the material.

106. A supplier quality engineer has received quarterly performance data for a key supplier. The supplier consistently meets delivery schedules but has been struggling with quality, as indicated by increasing defect rates. Using the following data, what is the engineer's best next step?

Metric	Quarter 1	Quarter 2	Quarter 3	Quarter 4
On-time delivery (%)	98%	97%	99%	96%
Defect rate (PPM)	200	300	400	600

a. Terminate the supplier due to the rising defect rate.

b. Implement a SCAR to address quality issues.

c. Maintain the supplier relationship since they meet delivery schedules.

d. Select a new supplier to mitigate future risks.

107. Which of the following tools would be the most appropriate for keeping management apprised of emerging technologies relating to the supply chain?

    a. SWOT analysis

    b. PEST analysis

    c. Portfolio analysis

    d. Risk analysis

108. Which of the following best describes an agreement with a supplier to reimburse a customer for failure of supplied goods/services?

    a. Warranty

    b. Guaranty

    c. Bond

    d. Certification

109. All of the following would be part of risk monitoring and review *except:*

    a. identifying new or upcoming risks.

    b. obtaining new information to update risk levels.

    c. assessing the controls in place to make sure they are working and appropriate.

    d. feasibility analysis.

110. Supplier development is an endeavor that adds value:

    a. by early involvement of the supplier quality group during the selection process.

    b. by partnering with a supplier to raise the level of quality and service performance.

    c. by developing strong personal relationships with the supplier's key contacts.

    d. by helping existing suppliers develop new core competencies.

111. Common elements of which of the following include a description of the product, its required characteristics, and any handling and packaging requirements?

    a. Specifications

    b. Control plans

    c. Capability analyses

    d. Sampling plans

112. Which of the following may be described as a tool that can be used to organize and display all of the knowledge and information that a team has regarding a particular problem?

    a. Matrix diagram

    b. Cause-and-effect diagram

    c. Force-field analysis

    d. Radar chart

113. Which of the following can be described as a situation where an individual's own interests could compromise their own judgment, decisions, or actions at work?

    a. Noncompliance

    b. Social responsibility

    c. Hoshin Kanri

    d. Conflict of interest

114. Which of the following is the stage of negotiation for discovery and diagnosis of the people, interests, options, and criteria?

    a. Planning

    b. Discussion

    c. Analysis

    d. Resolution

115. The costs associated with the operation and activities of the MRB are best characterized as:

    a. appraisal costs.

    b. prevention costs.

    c. internal failure costs.

    d. external failure costs.

116. ITAR regulations in the United States apply to all of the following *except*:

    a. weapons and ordinance.

    b. technical data related to defense items.

    c. basic system descriptions.

    d. spacecraft and associated equipment.

117. Who is the most responsible for risk?

    a. Supplier

    b. Customer

    c. Consumer

    d. End user

118. The notion of the "triple bottom line" of profits, the environment, and people comprises which of the following?

    a. TQM

    b. DEI

    c. Sustainability

    d. Compliance

119. Which of the following can be described as the practice of managing an organization so that it excels in all aspects and characteristics of products and services that are important to customers?

    a. Kaizen

    b. TQM

    c. Lean

    d. DMAIC

120. Which of the following is communicated to the auditee and should include audit objectives, scope, criteria, and identification of reference standards and documents?

    a. Audit schedule

    b. Audit report

    c. Audit record

    d. Audit plan

121. Leverage products are characterized as products with:

    a. low profit impact and low supply risk.

    b. low profit impact and high supply risk.

    c. high profit impact and low supply risk.

    d. high profit impact and high supply risk.

122. In which role in the RACI model do the individuals actively participate in an activity?

    a. Responsible

    b. Accountable

    c. Consulted

    d. Informed

123. The causal factor or factors that, if removed, will prevent the recurrence of the same situation describe:

    a. CAPA.

    b. FMEA.

    c. process control.

    d. root cause.

124. Postproduction risk management can be derived from various sources, including service personnel, training personnel, incident reports, and ______.

    a. production monitoring

    b. process validation

    c. customer feedback

    d. incoming inspection

125. The formal, cross-functional evaluation of a design to ensure that it meets requirements and to identify and resolve any problems is known as which of the following?

    a. Quality review

    b. FMEA

    c. Design review

    d. FTA

126. During a contract review for a new supplier, a supplier quality engineer identifies discrepancies between the technical specifications in the contract and the supplier's capability statements. What should the engineer focus on during the contract review process to prevent issues during production?

 a. Ensure the contract prioritizes the supplier's production lead times.

 b. Verify that the supplier's capabilities match the technical specifications in the contract.

 c. Confirm that the supplier has a valid ISO certification.

 d. Review the contract to ensure the pricing aligns with company budgets.

127. Suppliers of off-the-shelf components may initially be classified as what level of risk?

 a. Low

 b. Medium

 c. High

 d. Negligible

128. Regarding a measurement system, the following target represents which of the following?

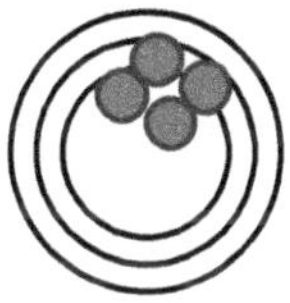

 a. Good accuracy and good precision

 b. Poor accuracy and good precision

 c. Good accuracy and poor precision

 d. Poor accuracy and poor precision

129. During which stage of the PDCA cycle would a variety of tools such as benchmarking and supplier scorecards be utilized?

 a. Plan

 b. Do

 c. Check

 d. Act

130. When using GD&T, MMC refers to:

a. median material condition.

b. mean material condition.

c. minimum material condition.

d. maximum material condition.

131. In-process acceptance is calculated by part acceptance rates. Apex Manufacturing, a supplier of key components, provided the following information for Q4: 864 parts were manufactured, of which 18 parts were found to be incorrect and had to be discarded. What is the in-process acceptance percentage for Q4?

a. 1.85

b. 2.08

c. 97.92

d. 98.15

132. The primary difference between supplier life cycle management (SLM) and supply chain management (SCM) is:

a. SLM's focus on supplier integration.

b. SLM's assessment of suppliers' assets and capabilities.

c. SCM's planning for turnover in suppliers.

d. SCM's focus on supplier integration.

133. The difference between the actual value of a characteristic and the measured value of that characteristic is referred to as ____________.

a. resolution

b. bias

c. measurement error

d. stability

134. Which of the following best describes the relationship between AQL and sample size?

a. As AQL increases, sample size increases.

b. They are not related; sample size is based only on lot size.

c. They are not related; sample size is based only on AQL.

d. As AQL increases, sample size decreases.

135. Which of the following is a step or place in a process in which a major or serious failure of a product can be introduced?

   a. Critical to quality

   b. Reaction plan

   c. Failure mode

   d. Critical control point

136. Sampling plans provide for effective and economical inspection. Although economical, sampling does provide a degree of risk to the:

   a. producer.

   b. consumer.

   c. regulators.

   d. producer and consumer.

137. Costs associated with customer complaints are best characterized as:

   a. appraisal costs.

   b. prevention costs.

   c. internal failure costs.

   d. external failure costs.

138. Which of the following is a form that is prepared for use in recording data?

   a. Histogram

   b. Pareto chart

   c. Check sheet

   d. Scatter diagram

139. A company may choose to assess a supplier's ability to incorporate technical changes in a timely manner as a way to judge the supplier's:

   a. ability to deliver on time.

   b. responsiveness to identified requirements.

   c. capacity to fulfill the company's orders.

   d. ability to continually improve costs.

140. The team phase where team members express their own opinions and ideas, often disagreeing with others, is referred to as:

   a. storming.

   b. performing.

   c. forming.

   d. norming.

141. A value indicating the relative risk of a potential failure is referred to as:

   a. failure.

   b. modality.

   c. RPN.

   d. severity.

142. Requirements in the technical review can include all of the following *except*:

   a. manufacturing environment (e.g., cleanroom).

   b. functional specifications.

   c. material properties.

   d. terms and conditions.

143. In PPAP, FAI stands for:

   a. final acceptance inspector.

   b. factory acceptance initiation.

   c. first article inspection.

   d. failure and inspection.

144. All of the following are examples of quality audits *except*:

   a. product audit.

   b. process audit.

   c. primary audit.

   d. software audit.

145. Before suppliers provide a quotation for a new project, what should the supplier consider?

a. Capability analysis

b. Capacity analysis

c. Feasibility analysis

d. Takt time analysis

146. All of the following are true regarding supplier-related organizational policies *except*:

a. they are best implemented at onboarding.

b. they protect corporate information.

c. they define requirements.

d. they are managed solely by the purchasing team.

147. The best way to determine which requirements to apply to which suppliers is to:

a. sort the suppliers by spend level.

b. prioritize suppliers that have been issued SCARs.

c. classify the suppliers according to risk.

d. prioritize the suppliers based on what they supply.

148. Which of the following models is the best for identifying stakeholders of a proposed change in a manufacturing process?

a. PDCA

b. DMAIC

c. SIPOC

d. DFSS

149. Bottleneck products are characterized as products with:

a. low profit impact and low supply risk.

b. low profit impact and high supply risk.

c. high profit impact and low supply risk.

d. high profit impact and high supply risk.

150. A purchasing system includes all of the following key activities *except*:

a. specification of requirements.

b. supplier selection.

c. supply chain management.

d. quality planning.

SOLUTIONS

1. c; Developing an emergency response plan and identifying alternative suppliers is the most effective way to mitigate the risk of a single-source supplier. This ensures supply continuity during potential disruptions. [II.A.2]

2. c; A clause that defines penalties for non-conformance ensures that the supplier is accountable for delivering products that meet the required quality and performance standards. [I.D.3]

3. d; None of choices (a), (b), or (c) are acceptable in an audit report. Auditors must strive for objectivity and to be factual. Since the report might be read by others, confidentiality (proprietary information) must be safeguarded. [V.A.2]

4. b; Implementing the solution is associated with the *do* phase of the Plan-Do-Check-Act (PDCA) cycle. This is where you carry out the actions that have been previously planned. [II.B.3]

5. a; The key word is "operation," indicating that we are looking at a process, not a product. System audits look at the relationships between different parts of the system. [III.B.2]

6. a; Increased globalization and outsourcing has given rise to the introduction of counterfeit materials and parts within the supply chain. Companies that sell counterfeit materials and parts because of loose supply chain oversight should face serious penalties and cannot excuse themselves by blaming their suppliers. [II.A.3]

7. b; The takt time to produce 1500 parts due in three days with one eight-hour shift per day is calculated by

$$\text{Takt time} = \frac{\text{Time available}}{\text{Number of units to be produced}}$$

$$= \frac{(8\text{ hours/day})(60\text{ minutes/hour})(3\text{ days})}{1500\text{ units}} = 0.96\text{ minutes per unit.}$$

   [IV.A.3]

8. a; Product specifications and engineering drawings are developed prior to any supplier selection steps. Inspection plans are used in ongoing production and are usually generated after the part qualification. [I.D.3]

9. d; *Groupthink* is a situation in which critical information is withheld from the team because individual members censor or restrain themselves, either because they believe their concerns are not worth discussing or because they are afraid of confrontation. [V.B.3]

10. d; (a) and (c) are criteria for selecting CTQs. Although checking a CTQ is usually desirable, that's the aftereffect of selecting the CTQ characteristic, not the criteria. [III.C.4]

11. a; Histograms are frequency distributions. Box plots provide information about the distribution of the data but do not explicitly provide frequencies. [VI.B.2]

12. c; Although (b) and (d) are also great answers and are true, at the evaluation level, (c) provides the true reasoning behind the mutual value of collaboration. [III.C.2]

13. b; Fair labor business practices are a key example in the social pillar, although they may also improve economic conditions. [VII.B.2]

14. d; Teams with new members often find themselves needing to re-form before accomplishing their objectives. [V.B.1]

15. a; Effective joint quality planning includes discussion and review regarding economic, managerial, and technological factors. [V.A.3]

16. a; Commitment and trust are essential to forming partnerships. [I.A]

17. c; Defects should be classified according to severity, which may be impacted by the other options. [IV.B]

18. c; Supplier C has elevated scores in both key risk indicators, representing the highest total risk to the company. The others should continue to be monitored. [V.C]

19. b; In addition to pricing, purchasing negotiates delivery schedules, payment terms, and other contract-related details, ensuring alignment between the supplier's capabilities and the company's needs. Product specifications and performance criteria are defined by engineering, and detailed analyses of a supplier's quality system would be conducted by supplier quality. Supplier corrective action plans may be eventually required, but this would be well after the supplier qualification stage. [I.D.1]

20. c; Scatter diagrams show the direction and strength of the relationship between two variables. Therefore, they can help identify relationships that should be explored further to determine if a cause-and-effect relationship exists. [VI.B.2]

21. c; Cycle time is the time it takes to complete a single task or all value-added activities within a process, while lead time measures the entire project, including any slack/waiting time. [III.B.1]

22. a; This is the definition of a first party audit. The other options involve audits being conducted by someone from outside the company. [V.A.1]

23. a; Although (b) and (c) are technically true, the real objective is to meet goals for quality, cost, and delivery; (d) has no bearing to this process. [III.A.2]

24. b; The house of quality is a graphical representation of these often competing and complex drivers. Demand planning determines how many parts we need. FMEA is a risk tool. The house of quality can be used to develop design inputs, not outputs. [III.A.1]

25. d; Process flow diagrams (PFDs) are the best option of those presented to identify internal and external customers of a process. An FMEA would describe how a customer may experience a failure but does not explicitly define all customers. [III.A.2]

26. d; Available data do not always provide a reliable basis for predicting the future. For unique types of risks or for new product types, historical data may not be available or there may be different interpretations of available data by different stakeholders (e.g., industry and regulators). [II.A.3]

27. a; A facilitator can turn team conflict into a problem-solving process by discussing conflict resolution techniques. [V.B.3]

28. b; *GMP* is a set of regulations regarding manufacturing; GLP is similar but for laboratory settings. [VII.B.1]

29. a; Working for two competing firms is a professional conflict of interest. Choosing a supplier that is owned by a manager's family member could be a relational conflict of interest. Advocating for a supplier in order to receive a signing bonus (commonly referred to as a "kickback") from that supplier is a financial conflict of interest. [VII.A]

30. c; Cpk will show all of the answers except (c). Fit and performance are design outputs, unrelated to process capability. [III.C.5]

31. a; Safety interlocks and alarms are effective methods to mitigate risk by preventing hazardous conditions or alerting users to danger. This helps ensure user safety when hazards cannot be eliminated. [II.A.3]

32. d; This definition of HACCP is paraphrased from the FDA definition. GLP involves controls in a laboratory setting, and the other options are regulations, not management systems. [VII.B.1]

33. d; Those from affective cultures have a greater tendency to show emotion in communications, while those from neutral cultures will tend to be more emotionally reserved. [VI.B.1]

34. c; Key data sources for analyzing historical performance include supplier corrective action requests (SCARs); corrective and preventive action (CAPA); internal nonconformances; historical capability; and results from regulatory, ISO, and customer inspections and audits. Planned or scheduled maintenance, age of equipment, historical downtime, first-pass yield, and rolled throughput yield are also great predictors of success. [I.C.3]

35. b; A Pareto chart can be used to help a process improvement team focus their efforts based on the PFMEA. [II.B.1]

36. b; Auditor notes are working materials and are private to the auditor, as the positions expressed in the notes may evolve during the audit. They are therefore not usually considered audit records. (A distinction must be made with regard to a regulatory inspection; e.g., the notes that FDA documents are part of the investigation. This is one important difference between audits and inspections.) [V.A.2]

37. b; Preferred suppliers have demonstrated consistent performance and may receive bidding priority. [I.B.3]

38. b; People from high-context cultures rely on contextual elements and situational factors to convey meaning. Therefore, it is important to clearly define details of requirements and specifications to avoid misinterpretation. [VI.B.1]

39. b; Of these steps, the first is to Develop a containment plan and implement and verify interim actions (D3), then Determine, identify, and verify root causes and escape points (D4), then Choose and verify permanent corrections (D5), and finally Implement and validate corrective actions (D6). [IV.C.2]

40. a; The marks are tightly grouped and on target, so they have good precision and good accuracy. [IV.C.1]

41. d; This is the definition of the critical path. The other terms are not generally used in project management. [V.A.5]

42. a; Both REACH and RoHS are environmental statutes regulating the import and use of hazardous materials. [VII.C.3]

43. a; Heijunka boxes or boards are visual scheduling tools used to improve the flow of production. They use cards that are arranged to represent the required timing of material flow. [IV.A.3]

44. d; Control charts detect the presence of assignable or *special* cause variation in a process. Run charts do not contain the control limits to reference, while scatter diagrams show the relationship between two variables and Pareto charts are frequency distributions. [VI.B.2]

45. c;

$$\text{OTD\%} = \frac{56\ (\text{orders delivered on time})}{58\ (\text{total orders delivered})} = 96.6\%.$$

[IV.A.1]

46. b;

$$\text{OTIF\%} = \frac{52\ (\text{orders delivered in full on time})}{58\ (\text{total orders delivered})} = 89.7\%.$$

[IV.A.1]

47. b; Supplier quality audits should generally be done prior to PPAP for specific parts. [III.C.4]

48. c; The best approach is to identify the root cause, correct it, and conduct a new FAI to ensure the product meets all specifications. The engineer will need to work with the supplier to accomplish this. [III.C.3]

49. a; The correct order is: Determine the scope of work; Determine the process and draft the initial document; Determine what tools are required; Determine the amount of time required to perform the task. [IV.A.3]

50. a; The effects lines cross inside the plot area; this plot shows the largest difference in slope between the two lines. [IV.C.1]

51. c; Policy disputes between a person reporting and a manager are likely not covered by whistleblower protection, as this would fall under the realm of opinion. [VII.C.3]

52. c; A team may have a facilitator who assists the lead auditor in organizing the team and making it more effective but does not participate as a team member. In the cross-functional team for a specific purpose and the long-term project team, the facilitator or coach is included as needed. The facilitator's role is to help the team resolve issues and reach its purpose or achieve its goals effectively. [V.B.2]

53. a; Real-time monitoring of quality metrics allows the engineer to detect issues as they arise and respond promptly, which is essential for effective risk monitoring. (b) and (c) are also useful but not as timely, as they lag indicators. [II.A.2]

54. d; An airbag malfunction in a car could lead directly to a severe customer injury; therefore, this defect would be classified as critical. [IV.B]

55. a; The control plan should list the steps to be taken when a process change has been detected. This serves as an aid to the responsible personnel during what is often a stressful time. The reaction plan section should cover requirements for containment and inspection of products suspected of having defects. It should also discuss disposition of parts found to be defective. Some control plans prescribe a more intense sampling protocol after certain corrective actions have been taken. [IV.C.2]

56. b; Supportive leadership style is characterized by low directive and high supportive behavior. It is suited for managing individuals who possess moderate competence and commitment. Such individuals are often described as the capable, but cautious, performer. They know their task but are not necessarily motivated about it. [VI.C]

57. c; The type of risk involved when a lot could be accepted when it does not conform to the specification is known as beta (β) risk. [II.B.2]

58. b; Audit team members may include qualified auditors, auditors in training, and technical experts. Team members accept assignments and report progress to the lead auditor or the team as a whole. Team members may be given responsibilities such as keeping records or taking meeting minutes. [V.B.2]

59. d; Transportation waste involves unnecessary movement of materials in production processes and can be indicated by the presence of large conveyor systems or large fleets of forklifts. [IV.A.3]

60. a; The purpose of quality function deployment (QFD) is to allow translation of customer expectations into specific engineering and quality characteristics. This may require multiple layers of QFDs. [III.B.2]

61. b; Trade secrets aren't officially protected but are widely unknown and generally require corporate espionage to uncover. [VII.C.2]

62. c;

$$\text{Acceptance \%} = \frac{33\ (\text{Lots accepted})}{35\ (\text{Lots received})} = 94.29\%.$$

[IV.A.1]

63. c; A *mission statement* explains the company's reason for existence. A *vision statement* is what or where the company would like to be in the future. [I.A]

64. b; The symbol ∥ indicates parallelism, which controls whether two features are geometrically parallel to each other within the specified acceptable tolerance. [III.C.2]

65. c; The ASQ Code of Ethics Fundamental Principles apply to ASQ members and ASQ certification holders. [VII.A]

66. a; An emotive communicator shows high degrees of dominance and sociability. This person typically displays action-oriented behavior, prefers informality, and is adept at persuasion. An example of an emotive communicator is an easygoing, down-to-earth quality manager who leads by example, has a friendly demeanor, and gets people to follow him or her without difficulty. [VI.B.1]

67. d; Registration comes first. Enforcement comes after. The World Intellectual Property Organization (WIPO) provides international protection through treaties (therefore, [b] is wrong), not registration. [VII.C.2]

68. d; Inventory levels should be based on production consumption rates. Lead time measures the time from ordering to fulfillment, while wait interval measures non-value-added waste. The on-time delivery (OTD) metric is important to monitor, as supply may be affected if the company is not performing to requirements, but it should not determine inventory levels. [IV.A.3]

69. b; Defining the frequency of required communications is the responsibility of the customer. [VI.A]

70. c; Deviations should only be acceptable if there are engineering studies demonstrating that the change will not have adverse or unintended consequences. [III.C.5]

71. d; Pareto charts and run charts may seem like a good option, but these charts do not include specifications, which could be included on a histogram. [III.B.1]

72. b; A Pareto chart is a quality tool that can be used to organize data for analyzing trends in categories of failures. [IV.A.1]

73. a; An effective supplier management program can reduce the risk involved when there are routine communications with the supplier. [II.B.2]

74. b; Surveys and interviews depend on the honesty and openness of trainees and do not assess skills, while multiple-choice tests may not work well for those with test anxiety. [V.B.3]

75. c; The C stands for *control* and is considered, along with a great problem statement, the most critical phase because many organizations make improvements but lose them over time to culture, entropy, employee behavior, or tribal knowledge. It has nothing to do with cost control. [V.A.4]

76. c; When focusing on criteria, the negotiator requires that the results are based on a specified objective standard. [VI.C]

77. d; The customer perspective defines an organization's value proposition and measures how effective the organization is in creating value for its customers through its goals, objectives, strategies, and processes. [I.C.2]

78. a; Supplier A is rated low in both criticality and spend level, representing the lowest level of risk to the company. The others should be monitored and periodically re-evaluated. [V.C]

79. d; (a) and (b) are incorrect. (c) is a part of *feedback* but not the best answer. Roles and responsibilities are key to a good relationship and smooth operation going forward. [VI.A]

80. b; This is the function of the Material Review Board (MRB), which is cross-functional in nature and may include representatives from other departments including engineering and quality. [IV.B]

81. d; The other options are previous stages in the customer–supplier relationship. [I.B.4]

82. c; Contingency planning (also called a plan B) is used for crisis management, business continuity, and asset security. [II.A.1]

83. b; Supplier-Input-Process-Output-Customer (SIPOC) diagrams are used to visualize a process flow; therefore, they are a good option for defining the scope of a process audit. [II.B.1]

84. a; ISO 9001 is an example of a quality management specification. International Traffic in Arms Regulations (ITAR) controls exports and imports of defense-related items and services; Registration, Evaluation, Authorization, and Restriction of Chemicals (REACH) and Restriction of Hazardous Substances (RoHS) restrict the use of hazardous chemicals and other substances. [V.C]

85. c; Supplier onboarding may also be referred to as supplier orientation. [VI.A]

86. b; Lead auditors are selected by the client or audit program manager. [V.B.2]

87. a; Although timelines and delivery are similar, these are timelines metrics. They deal with receipts of shipments. [V.A.3]

88. b; Unless they are a special supplier like an Active Pharmaceutical Ingredients (API) manufacturer, most suppliers will not be subject to regulatory inspections. Although remediation can and should be done up front when faced with poor audit results, (b) is a more comprehensive answer covering poor performance at any time or phase in the relationship. [V.A.4]

89. a; QMS maturity is weighted at 3, the highest of all the categories. [III.B.3]

90. c; C is the lowest at 24. [III.B.3]

91. c; Value stream mapping (VSM) is a type of process mapping that can be used to understand the flow of a process. Fault Tree Analysis (FTA) and cause-and-effect diagrams are root cause analysis tools, while control charts are used to monitor process performance over time. [IV.C.1]

92. d; Muda is another name for waste. [IV.A.3]

93. c; This is the definition of slack time. *Lead time* refers to the time required to complete an entire project, including slack or wait time. [V.A.5]

94. a; The supplier's responsibilities regarding receipt and storage of required materials should be included in the purchasing terms, not the quality agreement. [VII.C.1]

95. d; A clause that specifies the supplier's responsibility for resolving quality failures ensures that the root cause is identified and corrective actions are implemented, preventing future defects. It should also outline responsibilities for containment actions. (a) does not address defects; (c) does not cover actual failures. (b) is true but should be covered as part of the clause described in (d), which is the best answer of the choices given. [I.D.2]

96. b; Number 8 represents an undeveloped event. [II.B.1]

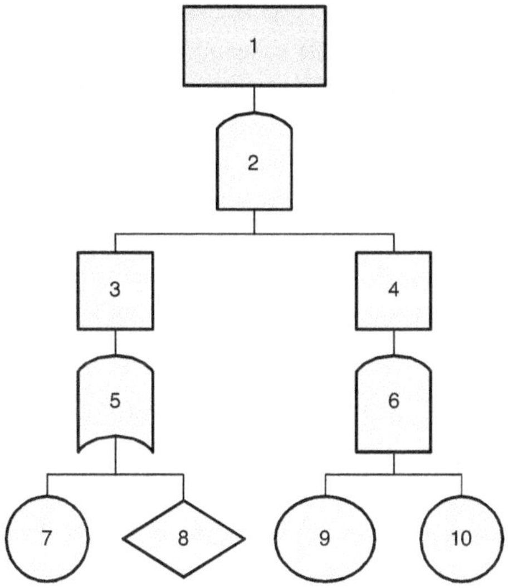

97. c; Row 3 and row 2 have the same Risk Priority Numbers (RPNs), but row 3 has a higher severity rating. [IV.C.2]

98. a; All the rows have the same RPN; therefore, look for the highest severity rating, then the highest occurrence. [IV.C.2]

99. b; This is the definition of stability with respect to a measurement system. [IV.C.1]

100. c; This is a definition of inclusion. [VI.B.1]

101. a; Validating the supplier's production process against control plans ensures that all requirements, especially CTQs, are met consistently. [III.C.5]

102. b; Although each answer is a good reason to handle an audit process well, the specific examples (communication and collaboration) indicate relationship management. [III.B.2]

103. a; REACH has been adopted to protect human health and the environment and to prevent risks posed by chemicals. [V.C]

104. c; There are three out-of-control conditions: Point 2 is higher than the upper control limit, points 11–17 are steadily decreasing, and points 20–27 are all on one side of the centerline without crossing. [IV.C.2]

105. c; The company should follow its formal, documented procedures. Each of these other scenarios would ideally be outlined in documented procedures, including when use of each is appropriate. [IV.B]

106. b; Implementing a Supplier Corrective Action Request (SCAR) is the best next step, as it addresses the supplier's rising defect rates while maintaining their reliable delivery performance. This gives the supplier an opportunity to improve before taking more drastic actions. [I.B.4]

107. a; A *strengths, weaknesses, opportunities, and threats (SWOT) analysis* is an effective strategic planning tool applicable to a business or project objective. Strengths and weaknesses are identified with respect to the internal capabilities of an organization, while opportunities and threats look outside the organization to identify opportunities for the organization and threats to the organization. [I.C.3]

108. a; This is a definition of warranty. The other options are guarantees of other parts of the contract or relationship. [I.D.2]

109. d; Risk monitoring and review consists of identifying new or upcoming risks, obtaining new information to update risk levels, and assessing the controls in place to make sure they are working and appropriate. Feasibility analysis should be done prior to entering into a supplier relationship. [II.A.2]

110. b; (a), (b), and (d) are all good answers, but (b) provides the most specific answer. [V.A.4]

111. a; Specifications include a description of the product, its required characteristics, and any handling and packaging requirements. Control plans and capability analyses refer to the specifications. Since sampling plans may be used for multiple products or materials, they generally do not refer to or contain any particular product specification. [V.C]

112. b; Cause-and-effect diagrams are effective tools for organizing and displaying all of the information that a team has regarding a particular problem. Matrix diagrams show the relationship between two items. Force-field analysis identifies forces that are present that support or work against the solution of a problem. Radar charts show the gap between current and ideal performance in multiple areas. [VI.B.2]

113. d; Conflict of interest is described as the presence of a situation where a person's own interests can compromise his or her judgment, decisions, or actions at work. [VII.A]

114. c; The analysis phase of negotiations is where the negotiator should seek to uncover and understand the underlying interests and issues on both sides. [VI.C]

115. c; Internal failure costs are costs incurred when a failure occurs in-house, and are usually associated with the cost of the management of nonconforming material, including scrap and rework. [I.C.1]

116. c; US Department of State International Traffic in Arms Regulations (ITAR) regulations describe the types of systems and items that are within the scope of the regulations. [V.C]

117. b; Although a supplier may conduct risk management activities, the customer is ultimately responsible for risk. [II.A.1]

118. c; Sustainability is concerned with balancing the *triple bottom line* of profits, the environment, and people. [VII.B.2]

119. b; This is a definition of Total Quality Management, or TQM. The other choices are approaches that may be used to support an overall TQM effort. [VI.C]

120. d; The audit schedule should be included in the audit plan, while an audit report is completed after the audit and forms part of the audit record. [V.A.1]

121. c; Leverage products are those with high profit impact and low supply risk. Purchasing approaches to consider here include using your full purchasing power, substituting products or suppliers, and placing high-volume orders. [II.A.4]

122. a; The RACI (responsible, accountable, consulted, and informed) matrix is usually defined in detail by a team or committee, and in some cases by an individual. The roles are:

    - *Responsible.* Individuals who actively participate in an activity.
    - *Accountable.* The individual ultimately accountable for results. Only one individual may be accountable at a time.
    - *Consulted.* Individuals who must be consulted before a decision is made
    - *Informed.* Individuals who must be informed about a decision because they are affected. These individuals do not need to take part in the decision-making process. [VI.C]

123. d; Solving a process problem means identifying the root cause and eliminating it. The ultimate test of whether the root cause has been eliminated is the ability to toggle the problem on and off by removing and reintroducing the root cause. [IV.C.1]

124. c; Postproduction information can be part of established QMS procedures (e.g., monitoring and measuring). Manufacturers must establish procedures to collect information from various sources, such as service personnel, training personnel, incident reports, and customer feedback. The other choices are related to preproduction or production activities. [II.A.2]

125. c; This is the definition of a design review. A quality review may be conducted later after the design is finalized to ensure that production processes are capable of producing product that meets specifications that were finalized in the design review. [III.A.1]

126. b; Verifying that the supplier's capabilities align with the technical specifications ensures that the supplier can meet product requirements, preventing potential production issues. [I.D.4]

127. a; Off-the-shelf components can usually be classified as low risk. [IV.A.2]

128. b; The marks are tightly grouped but off target, so they have good precision but poor accuracy. [IV.C.1]

129. c; The *check* stage is where a team will want to verify that actions are producing the desired results. During the check stage, various tools such as benchmarking and supplier scorecards are utilized. [I.B.2]

130. d; MMC indicates maximum material condition, e.g., the smallest hole or the largest diameter allowed. [III.C.2]

131. c;

$$\frac{864\ (\text{total parts}) - 18\ (\text{rejected parts})}{864\ (\text{total parts})}.$$

[IV.A.1]

132. d; SLM is an integrated approach that considers business and quality needs of the organization. [I.B.1]

133. c; Measurement error may occur in any direction, while bias is systematic error in a particular direction. All bias is error, but not all error is bias. [III.C.3]

134. d; AQL and sample size are inversely related. AQL and lot size are both required to determine the appropriate sample size. [III.C.3]

135. d; This is the definition of a critical control point. Critical control points are fundamental to HACCP. They may also be included in CTQs, but CTQ is a broader concept than what is being described in this question. [III.C.3]

136. d; Sampling plans provide for effective and economical inspection. Although economical, sampling does provide a degree of risk to the producer and consumer. [II.B.2]

137. d; External failure costs are costs incurred when a failure occurs while the customer owns the product; these may also be referred to as escapes. [I.C.1]

138. c; Check sheets are the basic quality tool used to collect data, while the other options are tools used to display that data. [IV.C.1]

139. b; A supplier's responsiveness to identified requirements indicates how quickly and how well the supplier responds to communication regarding changes in specifications and requirements. [III.B.1]

140. a; *Team storming* is when members express their own opinions and ideas, often disagreeing with others. [V.B.1]

141. c; The *Risk Priority Number (RPN)* is a value indicating the relative risk of the potential failure. RPN is the product of the severity, probability of occurrence, and probability of detection. [II.B.1]

142. d; All of the items (a), (b), and (c) should be included. Terms and conditions should be included in a contract review, not a technical review. [III.C.1]

143. c; FAI is the abbreviation for first article inspection, the systematic layout and 100% verification of the part to the engineering drawing, an anchor in PPAP. [III.C.4]

144. c; Product audits, process audits, and software audits are all common types of quality audits. [V.A.1]

145. c; Before suppliers provide a quotation for a new project, the supplier should consider performing a feasibility analysis. Capability analysis, capacity analysis, and takt time analysis are elements of a feasibility analysis. [I.E]

146. d; Supplier-related organizational policies are best implemented at onboarding, help protect corporate information, and define requirements. They are best managed by a cross-functional team. [VII.C.1]

147. c; Suppliers should be classified according to risk. The other options may be individual factors in risk classification, along with other factors. [IV.A.2]

148. c; SIPOC can be used to identify stakeholders. [I.B.1]

149. b; Bottleneck products are those with low profit impact and high supply risk. Useful approaches here include over-ordering when the item is available (lack of reliable availability is one of the most common reasons that supply is disrupted) and looking for ways to control vendors. [II.A.4]

150. d; Quality planning activities may also include purchasing activities but is not limited to such. [III.A.2]

Appendix
The ASQ Certified Supplier Quality Professional Body of Knowledge (2023)

The topics in this Body of Knowledge include subtext explanations and the cognitive level at which the questions will be written. This information will provide useful guidance for both the Exam Development Committee and the candidate preparing to take the exam. The subtext is not intended to limit the subject matter or be all-inclusive of that material that will be covered in the exam. It is meant to clarify the type of content that will be included on the exam. The descriptor in parentheses at the end of each entry refers to the maximum cognitive level at which the topic will be tested. A complete description of cognitive levels is provided at the end of this document.

I. Supplier Strategy (20 Questions)

A. Supply Chain Vision/Mission

Assist in the development and communication of the supply chain vision/mission statement. (Apply)

B. Supplier Lifecycle Management

1. Supplier Selection

Develop the process for supplier selection and qualification, including the identification of sub-tier suppliers using tools such as SIPOC, decision analysis, and total risk factor analysis. (Create)

2. Performance Monitoring

Develop the supplier performance monitoring system, including expected levels of performance, process reviews, performance evaluations, improvement plans, and exit strategies. (Create)

3. Supplier Classification System

Define and develop a supplier classification system (e.g., non-approved, conditionally approved, approved, preferred, certified, partnership, and disqualified). (Create)

4. Partnerships and Alliances

Identify and analyze strategies for developing customer-supplier partnerships and alliances. (Analyze)

C. Supply Chain Cost Analysis

1. Cost Reduction

Identify and apply relevant inputs to prioritize cost reduction opportunities. (Analyze)

2. Supply Chain Rationalization

Interpret and analyze the optimization of a supply base to improve spending and leverage investments into supplier quality or risk reduction. (Analyze)

3. Make/Buy Decisions

Provide input on make/buy decisions using internal and external capability analysis. Apply tools such as SWOT analysis and use historical performance to analyze requirements. (Analyze)

D. Supplier Agreements or Contracts

1. Terms and Conditions

Review and provide input for developing terms and conditions that govern supplier relationships to ensure quality considerations are addressed. (Apply)

2. Supplier Agreements

Identify elements of supplier agreements (e.g., business and legal approach/requirements). (Understand)

3. Quality Agreements

Analyze the elements of quality agreements/requirements (e.g., other levels of approval/review). (Analyze)

4. Finalization Controls

Describe controls used to finalize terms and conditions that govern supplier relationships (e.g., agreements, contracts, and purchase orders). (Understand)

E. Deployment of Strategy and Expectations

Communicate strategy internally and communicate expectations to suppliers externally. (Apply)

II. Risk Management (19 Questions)

A. Strategy

1. System

Develop a risk-based approach to manage the supply base, including business continuity, contingency planning, and supply chain resilience. (Create)

2. Product/Service Risk Mitigation

Develop and implement a risk mitigation plan for predicting, minimizing, monitoring, and/or controlling risks. (Create)

3. Prevention Strategies

Identify and evaluate strategies and techniques such as supply chain mapping, avoidance, detection, and mitigation used to prevent the introduction of counterfeit parts, materials, and services. (Evaluate)

4. Supplier Risk Identification and Categorization

Identify supplier risks and develop categorization (e.g., organizational, business, security, and product) using tools and models, such as the Kraljic portfolio segmentation model. (Create)

B. Analysis and Mitigation

1. Analysis

Identify, assess, and prioritize risks to supplier quality using tools such as decision analysis (DA), failure mode and effects analysis (FMEA), fault tree analysis (FTA), and process auditing. (Evaluate).

2. Mitigation Control

Develop and deploy controls such as inspection and test plans. Prioritize mitigation activities and sustain a risk mitigation plan appropriate to the risk of the product/service. (Create)

3. Mitigation Effectiveness

Verify the effectiveness of the control plan and improve, if necessary, using continuous improvement methods such as plan-do-check-act (PDCA), lean, and product auditing tools. (Create)

III. Supplier Selection and Part Qualification (27 Questions)

A. Product/Service Requirements Definition

1. Internal Design Reviews

Identify and apply common elements of the design review process, including roles and responsibilities of the participants. (Apply)

2. Identifying Requirements

Identify and apply internal requirements (e.g., interrelated functional business units) for product or service in collaboration with stakeholders, including the requirements for supply chain, sub- tier suppliers, and manufacturability evaluation. (Evaluate)

B. Supplier Selection Planning

1. Supplier Comparison

Evaluate existing suppliers' and distributors' capabilities, capacities, past quality, delivery, price, lead times, and responsiveness against identified requirements. (Evaluate)

2. Potential Suppliers Evaluation

Assess potential new suppliers against identified requirements using tools such as self- assessments, audits, financial analysis, and quality function deployment. Verify third-party certification status and regulatory compliance and analyze and report on results of assessments to support the supplier selection process. (Evaluate)

3. Supplier Selection

Evaluate and select suppliers based on analysis of assessment reports and existing supplier evaluations using decision analysis tools such as weighted decision matrices and selection matrices. (Evaluate)

C. Part, Process, and Service Qualification

1. Technical Review

Interpret and evaluate technical specification requirements and characteristics such as views, title blocks, dimensioning and tolerancing, and apply GD&T symbols as they relate to the product and process. (Evaluate)

2. Supplier Relations

Collaborate with suppliers to define, interpret, and classify quality characteristics for the part/process/service. (Evaluate)

3. Process and Service Qualification Planning

Develop a part/process/service (e.g., calibration, laboratory, software, and design) qualification plan with supplier and internal team that includes service provider audit, calibration requirements, sample size, first article inspection, measurement system analysis (MSA), process flow diagram (PFD), failure mode and effects analysis (FMEA), control plans, critical to quality (CTQ), inspection planning, capability studies, material and performance testing, appearance approval, and internal process validation. (Analyze)

4. Part Approval

Understand production part approval process (PPAP) requirements and ensure suppliers understand the processes required to produce parts with consistent quality during an actual production run at production rates. (Understand)

5. Validate Requirements

Collaborate with internal team to interpret the results of the executed qualification plan for the part/process/service, including reviewing Certificate of Compliance (CoC), Certificate of Analysis (CoA), and production readiness reviews (PRR). (Evaluate)

IV. Supplier Performance Monitoring and Improvement (29 Questions)

A. Supplier Performance Monitoring

1. Supplier Metrics

Define, implement, and monitor supplier performance metrics such as quality, delivery (e.g., on- time delivery [OTD] and on-time in full delivery [OTIF]), cost, and responsiveness. (Evaluate)

2. Supplier Performance

Analyze supplier performance data (e.g., warranty analysis/field returns and defect rates) and develop periodic reports (e.g., scorecard and dashboards). (Analyze)

3. Supplier Process Performance

Define and implement lean principles and applications such as 5S, kaizen, value stream mapping, supplier process capabilities and controls, 8 wastes, single minute exchange of dies (SMED), kanban, muda, standardized work, takt time, and error-proofing to reduce waste and increase performance. (Evaluate)

B. Assess Nonconforming Product/Process/Service

Segregate, control, and evaluate nonconforming materials to determine whether a material review board (MRB) requires disposition. Conduct risk assessments to prevent future discrepancies. (Evaluate)

C. Supplier Corrective and Preventive Action (CAPA)

1. Root Cause Analysis Tools and Methods

Evaluate the root cause analysis of a problem using tools such as cause and effect diagrams, Pareto analysis, 5 Why's, fault tree analysis, design of experiments (DOE), brainstorming, check sheets, measurement system analysis (MSA), production records, and review of process flow. (Evaluate)

2. Collaboration with Supplier

Evaluate and implement supplier corrective/preventive action and review its effectiveness and robustness with supplier. Understand the process of updating failure mode and effects analysis (FMEA) and process control plan, and understand statistical process control (SPC), 8D, and product/process design change. (Evaluate)

V. Supplier Quality Management (26 Questions)

A. Supplier Quality Monitoring

1. Supplier Audit

Apply the stages of a quality audit, including audit planning, conducting the initial audit, and executing periodic reevaluation. Understand and apply the various types of quality audits (e.g., product, process, and management system) and audit methods (e.g., virtual, on-site, and desktop). (Apply)

2. Audit Reporting and Follow-up

Apply and analyze audit reporting and follow up, including verification of the effectiveness of corrective action. (Analyze)

3. Supplier Communication

Evaluate various communication techniques such as periodic reviews, metric and performance indices, change management, notifications, recalls, change requests, and business updates.

Maintain active communication with suppliers to assess risk and take appropriate action. (Evaluate)

4. Supplier Development and Remediation

Identify and analyze present and future training needs and gaps, using quality methods and tools such as kaizen and benchmarking. Use process improvement tools such as DMAIC, cycle time reduction, defect rate, and cost reduction. Evaluate supplier remediation to develop and manage improvement plans. (Evaluate)

5. Project Management Basics

Understand and apply various types of project reviews, such as phase-end, management, and retrospectives or post-project reviews to assess project performance and status, to review issues and risks, and discover and capture lessons learned from the project. Apply forecasts, resources, schedules, and task and cost estimates to develop and monitor project plans. (Apply)

B. Teams and Team Processes

1. Team Development

Identify and describe the various types of teams and the classic stages of team development: forming, storming, norming, performing, and adjourning. (Apply)

2. Team Roles

Define and describe various team roles and responsibilities for leader, facilitator, coach, and individual member. (Understand)

3. Performance and Evaluation

Describe various techniques to evaluate training, including evaluation planning, feedback surveys, pre-training testing, and post-training testing. (Understand)

C. Compliance with Requirement and Supplier Categorization

Understand and evaluate compliance with regulations and industry standards (e.g., RoHS, Governmental regulatory authorities, and ISO), specifications, contracts, agreements, and certification authority. Evaluate and categorize suppliers based on risk and performance. (Evaluate)

VI. Relationship Management (16 Questions)

A. Supplier Onboarding

Understand and apply processes for orientation of suppliers such as providing overview of company, vision, mission, guiding principles, overall requirements, expectations, and criticality of product, service, and delivery requirements. (Apply)

B. Communication

1. Techniques and Mediation

Identify and apply communication techniques (e.g., oral, written, and presentation) specifically for internal stakeholders and suppliers to resolve issues. Apply different techniques when working in multi-cultural environments. Identify and describe the impact that culture, communications, and Diversity, Equity, and Inclusion (DEI) can have on an organization. (Evaluate)

2. Reporting Using Quality Tools

Use appropriate technical and managerial reporting techniques for effective presentation and reporting, including the seven classic quality tools: Pareto charts, cause and effect diagrams, flowcharts, control charts, check sheets, scatter diagrams, and histograms. (Analyze)

C. Leadership and Collaboration

Understand and apply techniques for coaching suppliers through regular communications, influencing without authority, negotiation techniques, conflict resolution techniques, and establish clear roles and responsibilities of internal stakeholders and suppliers using tools such as a RACI matrix (responsible, accountable, consulted, and informed). (Evaluate)

VII. Business Governance, Ethics, and Compliance (13 Questions)

A. ASQ Code of Ethics

Determine appropriate behavior in situations requiring ethical decisions, including identifying conflicts of interest, and recognizing and resolving ethical issues. (Apply)

B. Compliance and Sustainability

1. Compliance

Understand issues of compliance and their applicable policies, laws, and regulations (e.g., conflict of interest, confidentiality, and bribery). (Apply)

2. Sustainability

Understand and recognize the importance of environmental, social, and governance factors and adhere to applicable sustainability policies. (Understand)

C. Confidentiality

1. Organizational Policies

Apply organizational policies for executing appropriate agreements such as non-disclosure, quality, and change notification agreements. (Apply)

2. Intellectual Property

Apply procedures for protecting the intellectual property of an organization and its suppliers. (Apply)

3. Illegal Activity

Understand and interpret policies for reporting observations and deviations that could be perceived as illegal activity. (Apply)

LEVELS OF COGNITION BASED ON BLOOM'S TAXONOMY—REVISED (2001)

In addition to **content** specifics, the subtext for each topic in this BOK also indicates the intended **complexity level** of the test questions for that topic. These levels are based on "Levels of Cognition" (from Bloom's Taxonomy—Revised, 2001) and are presented below in rank order, from least complex to most complex.

Remember

Recall or recognize terms, definitions, facts, ideas, materials, patterns, sequences, methods, principles, etc.

Understand

Read and understand descriptions, communications, reports, tables, diagrams, directions, regulations, etc.

Apply

Know when and how to use ideas, procedures, methods, formulas, principles, theories, etc.

Analyze

Break down information into its constituent parts and recognize their relationship to one another and how they are organized; identify sublevel factors or salient data from a complex scenario.

Evaluate

Make judgments about the value of proposed ideas, solutions, etc., by comparing the proposal to specific criteria or standards.

Create

Put parts or elements together in such a way as to reveal a pattern or structure not clearly there before; identify which data or information from a complex set is appropriate to examine further or from which supported conclusions can be drawn.

www.ingramcontent.com/pod-product-compliance
Lightning Source LLC
Chambersburg PA
CBHW081127180926
51611CB00005B/70
9781636941585